D0385356

THE UNIQUENESS OF THE INDIVIDUAL

THE UNIQUENESS OF THE INDIVIDUAL

by

P. B. MEDAWAR

formerly Jodrell Professor of Zoology and Comparative Anatomy, University College, University of London, and Mason Professor of Zoology in the University of Birmingham

Second Revised Edition

Dover Publications, Inc.

New York

Published in Canada by General Publishing Company, Ltd., 30 Lesmill Road, Don Mills, Toronto, Ontario.
Published in the United Kingdom by Constable and Company, Ltd., 10 Orange Street, London WC2H 7EG.

This Dover edition, first published in 1981, is a revised edition, with a new Introduction by the author, of the work originally published by Methuen & Co, Ltd, London, in 1957.

International Standard Book Number: 0-486-24042-8
Library of Congress Catalog Card Number: 80-66840

Manufactured in the United States of America
Dover Publications, Inc.
180 Varick Street
New York, N.Y. 10014

Contents

Introduction *page* vii

1. Old Age and Natural Death (1946) 1
2. An Unsolved Problem of Biology (1952) 28
3. A Note on 'The Scientific Method' (1949) 55
4. A Commentary on Lamarckism (1953) 63
5. The Pattern of Organic Growth and Transformation (1954) 88
6. The Imperfections of Man (1955) 102
7. Tradition: The Evidence of Biology (1953) 114
8. The Uniqueness of the Individual (1956) 123

Index 157

Acknowledgements

I acknowledge my gratitude to their respective publishers for permission to reproduce the following articles: Old Age and Natural Death: *Modern Quarterly*, vol. 1 (1946), p. 30. An Unsolved Problem of Biology: H. K. Lewis, London, 1952. A Note on 'The Scientific Method': *Nineteenth Century* (1949), p. 115. A Commentary on Lamarckism: *Bull. Nat. Inst. Sci. India*, vol. 7 (1953), p. 127. The Pattern of Organic Growth and Transformation: *The Times Science Review*, no. 12 (1954), p. 10. The Imperfections of Man: *Twentieth Century* (1955), p. 236.

Introduction to the Dover Edition

The republication of a collection of essays written by a young and little known biologist between twenty and forty years ago might easily be construed as arrogance on the author's part—and as grave imprudence on the publisher's. But a case can be made for the enterprise, and it is this: for anyone interested in the history of ideas (few will admit being bored by it) these essays give a cross-sectional view of problems of the kind that interested young biologists on the eve of the molecular revolution.

Young biologists today may be amused and I suspect a little contemptuous upon learning how great a part the theory of evolution still played in people's thoughts. No less than five of these essays are upon evolutionary themes. Surely the theory of evolution could have been taken to be thoroughly well established at the time these essays were written? Certainly the theory of evolution had achieved a degree of general acceptance that would have made it an act of conscious eccentricity to disavow it. We were all Darwinists, too—all, that is to say, except for a small group of Lamarckists who regarded themselves as, and were, a persecuted minority. But Darwinism in those days was far gone in what William Bateson in his classic treatise on Mendel's *Principles of Heredity* (1909) called 'the apathy characteristic of an age of faith'. Darwinism meant that evolution was thought to proceed by 'the action of natural selection upon variation in every direction'. But this glib formula—which I quote from the writings of one of the leading professors of zoology of his day—was beginning to lose its power to persuade. Nowhere is that dissatisfaction more clearly to be seen than in that great baroque masterpiece of biological literature, D'Arcy Thompson's essay *On Growth and Form.*[1]

Biology was in fact on the eve of a revolution as important as the molecular revolution that was yet to come: this was the genetical revolution in evolutionary theory; that is to say, the refounding of evolutionary theory on the basis of Mendelian genetics. Sewall Wright[2] was the prime mover in this new style of thinking, to the theory of which important and distinctive contributions were made by J. B. S. Haldane, R. A. Fisher and Alan Robertson, though the names of H. T. J. Norton and A. J. Lotka are by no means to be forgotten. At first these endeavours were purely theoretical, but in due course the new notions of gene-frequency equilibria and the supplanting of one allele by another were tested in the field by Theodosius Dobzhansky, Ernst Mayr and the Oxford school of ecological geneticists—especially Cyril Clarke, Arthur Cain and Philip Sheppard—and found to have high explanatory value. Microbial systems, too, have lent themselves readily to the construction of models of the evolutionary process reconceived in terms of gene frequencies. Accordingly we hear fewer derisory remarks about 'bean bag genetics' nowadays. When I wrote these essays, however, old-fashioned no-nonsense zoologists still had plenty of grounds upon which to snipe at the theorists, whose work they didn't really understand.

One of the most popular misconceptions about the theory of evolution by natural selection is that which treats it as the *denouement* of the following train of thought: (a) organisms produce offspring in numbers vastly in excess of their needs; (b) only a minority survive; therefore (c) only those survive which are best equipped to do so, the 'fittest'. The catch in this Malthusian syllogism, pointed out years ago by R. A. Fisher, lies in its major premise (a). So far from producing a vastly excessive number of offspring, most organisms produce just about that number which is sufficient and necessary to perpetuate their kind. Degree of fecundity is one of the consequences of natural selection: it is not the cause. Nidicolous birds, Lack has shown us, illustrate this truth with particular clarity; they do in fact lay clutches of a certain size, though they could lay more eggs (the egg industry is founded upon the inexhaustible gullibility of the domestic fowl) and they could, of course, lay fewer. Having regard to all the exigencies of giving birth to and rearing eggs and young, the size of its clutch is just about that which gives each species its greatest likelihood of self-perpetuation.

A second misconception may be aptly called the Zenonian, because of a certain family likeness to the argument which purports to show that Achilles can never overtake the tortoise, nor an arrow reach its target. Any substantial adaptation, it was argued, can only be achieved by the adding up, over very many generations, of single all-but-infinitesimal adaptive changes which, being of 'inappreciable' advantage to their owners, offered nothing for selection to get to grips upon. Luckily, selection does not go by human assessments of its efficacy; it can be shown that even so slight a selective advantage as that which allows 1001 possessors of the gene to perpetuate themselves for every thousand that lack it—a selective advantage of only 0.1 per cent.—may eventually cause it to prevail.

Philosophers have sometimes beguiled themselves with a third argument allegedly discreditable to Darwinism. Selection (the argument runs) can select only from what is available and 'given', i.e., can select only from within the compass of existing variation and diversity; by selection, all men or all deer could be caused to become as tall or swift as the tallest or swiftest now among them, but not still taller nor swifter still. The error here was to equate all variation with overt variation, to suppose that all genetic goods are always in the shop window;

in reality, though, the genetical mechanism is such that there are deep resources of hidden variation, of possible animals only awaiting the occasion to become real. If mutation were to take a holiday, as Death did in Casella's play, evolution would not come to a standstill, though in the long term the range of variation would be sadly confined. One of the aspects of Darwinian theory that many old-fashioned zoologists found most repugnant was the idea that genetic variations do not arise in response to an organism's needs and do not, except by accident, gratify them. For this reason the metaphysical theory still widely known as 'Lamarckism' still claimed a number of adherents; at this distance of time the names that have remained in my mind are those of H. Graham Cannon, W. McDougall and E. W. MacBride. Its psychological appeal is self-evident: in 'cultural' or 'exogenetic' heredity, mediated through tradition and indoctrination and learning, a Lamarckian process is clearly at work. Moreover it gratifies our sense of the fitness of things that our aspirations and endeavours can somehow be credited to our offspring in the form of an inheritance of special capabilities. Lamarckism, moreover, has a powerful political appeal,[3] for the working of Darwinism depends upon the premise that organisms are born unequal, whereas Lamarckian heredity can be reconciled with the political notion that all men are born equal—a political but not genetic truth—and that what they become is what the environment makes of them.

These considerations seem to me to justify my writing an extended analysis (*A Commentary on Lamarckism*) of various shades of opinion passing under the general name of Lamarckism.

The Lamarckian interpretation offers special temptations to amateurs of biology (the heading under which many of the then-champions of Lamarckism must be classified). I think I was right to say that the class of adaptation that offers the greatest temptation to Lamarckism was that which in one essay I designated as 'Class B'. Adaptations so classified have the property that, although developmentally 'programmed'—as we now say,—they *could* have arisen in an individual's own lifetime under the influence of use; I had in mind palmar and other flexure lines and the pattern of articulation of joints. Modern readers will be amused to see that the interpretation of bacterial adaptation to, for example, antibiotics or new substrates, whether Darwinian or Lamarckian in character, is still described as 'controversial'. It was so because an extremely distinguished physical chemist, C. N. Hinshelwood, who became President of the Royal Society of London and won the Nobel Prize for Chemistry in 1965, insisted over a period of many years that bacterial adaptation was the outcome of a Lamarckian process, i.e., that the adaptation of the population as a whole was the outcome of a hereditary adaptation of the individual members of the population, transmitted to their progeny.

Everybody recognises as one of the great achievements of modern biology the successful recruitment of a number of physicists and chemists who inaugurated the molecular revolution and brought added strength to biology in many ways. Hinshelwood was not, as he could conceivably have been, the first of these because, holding as I believe a somewhat élitist view of the stature of the physical sciences, he did not take enough trouble to acquaint himself with his subject matter at a biological level.

It was I when a young teacher at Oxford, who—with some trepidation and sense of daring—made known to Hinshelwood the existence in yeasts of a Mendelian genetical process controlling the ability of yeast cultures to metabolize a number of simple sugars: I truly believe this information was something of a revelation to Hinshelwood, but in spite of the prima facie likelihood it established of the essential rightness of a Darwinian interpretation of bacterial adaptation, it did not cause him to change his mind.

This book contains the only available exposition of my theory of the evolution of ageing (the essays *Old Age and Natural Death* and *An Unsolved Problem of Biology*). I think these essays must have caused some surprise at the time because in spite of the gravity of their subject matter I did not feel, and have not since felt, any obligation to pull a long face on the subject.

The theory I propounded was in the new style of genetical reasoning, i.e., was in the style of the newly emerging genetical theory of natural selection; J. B. S. Haldane liked it and I was flattered when R. A. Fisher—never given to squandering praise—said he thought my approach 'admirable'.

Although the argument seemed to me then—and seems to me still—self-evidently true, I was a bit uneasy about it, so I was therefore relieved rather than put out when a distinguished evolutionary biologist, G. C. Williams, propounded independently an essentially similar theory.[4]

By choosing an inorganic model—a microcosm of test tubes (p. 43)—to illustrate my theory I had hoped to escape any imputation that my reasoning was circular, as August Weismann's clearly was. Nevertheless Kirkwood and Holliday[5] have expressed misgivings about the matter, suggesting that unless a process of deterioration with increasing age is already in progress a gene cannot 'tell the time' and thus come into action earlier or later as the case may be—an essential element of Williams's and my theory. I do not regard this criticism as wholly just because senescent deterioration in the course of a lifetime is not the only means by which a gene might 'tell the time'.

Until the delivery of Dr D. A. Pyke's Claude Bernard Lecture[6] I had believed that diabetes might be an example of the genetically enforced postponement of the action of deleterious genes—a prediction of Williams's and my theory. I had in mind the difference between insulin-dependent diabetes of juvenile origin and the non insulin-dependent diabetes of middle life. The latter, however, cannot possibly represent a postponement into later life of the former because the two are quite different diseases with quite distinct genetic determinations. Juvenile diabetes is HLA-associated (i.e., associated with genetical differences of the kind that influence the outcome of organ transplantations) but is without much evidence of concordance between twins, whereas, as Pyke's lengthy and thorough studies showed, non insulin-dependent diabetes is characterized by a high degree of twin concordance and is associated, as insulin-dependent diabetes is not, with the strange and perhaps etiologically significant phenomenon of chlorpropamide-induced alcohol flushing.

Research on ageing, like its subject matter, does not move very fast, so a comparatively small number of glosses will bring the existing text up to date. In a recent symposium on the subject[7] I called attention to a number of shortcomings in almost universally used actuarial measurement of the degree of senescence in terms of the force of mortality: all kinds of dangers may grow out of the practice of recording the progress of a continuous process in terms of a frequency distribution of a single event, viz., death; moreover some ageing processes are not accompanied by an increase of vulnerability such as would show up on the actuaries' life table. The menopause, I point out, is one.

In a general discussion of the definition and measurement of ageing[7] I pointed out that the bringing forward in life of the time of action of 'good' genes would, if my reasoning were correct, come to a standstill at the time of life at which an individual's reproductive value was at its highest; I take this opportunity to mention that A. J. Coale[8] gave a much more polished account than mine of the demographic properties of a potentially immortal population.

Turning away now from the evolutionary argument, I must in retrospect say that I should not have been so critical of Weismann's notion of 'an inherent limitation in the power of germ cells to divide' if I had been able to foresee Leonard Hayflick's now well-known demonstration[9] that—contrary to the opinion we had all taken for granted—the lifespan of diploid tissue cultures is determinate and comes to an end after a finite number of cell divisions that depends amongst other things upon the age of the individual from which the tissues were taken for culturing. The same applies to Lansing's equally well-known work on the determinancy of the lifespan of the rotifer *Philodina citrina*—rotifers being highly neotenous organisms with a small fixed number of cells, which look and behave as we should expect sexually mature annelid-type embryos to behave. In *P. citrina* there are thirty-two oocytes in the adult—the product of five successive cell divisions.[10] It is specially interesting that all clones of rotifers pass through the same number of cell divisions in the same number of generations and that clones propagated through a lineage of older mothers die out much more quickly than clones propagated through lineages of younger mothers. Only a modern Francis Galton would have the time and enthusiasm to find out if the same principle might not apply to human beings or other mammals: there must be many families in which records of longevity and the ages of mothers at the birth of their children are known over a period of many generations. It would be profoundly interesting to know if longevity were any less in lineages that by chance went through older mothers than in lineages propagated through younger mothers. The question is specially relevant because human ovaries contain a finite number of egg cells, liberated at regular intervals through life; a line of descent passing through older mothers is thus analogous to those rotifer lineages that are propagated through older eggs. Nevertheless, whatever the behaviour of rotifers and whatever may turn out to be the truth about human beings, Weismann's notion that the cell might divide 10,000 or 100,000 times before suddenly stopping had that madly nonquantitative character which at one time used to be thought characteristic of biologists. It was not characteristic of Charles Minot, though, because his notion that ageing starts very early in life is upheld by the consideration that all 'growth curves' that are used to graduate human growth data (i.e., represent size as a function of age) have the property that the specific acceleration of growth, that is to say the quantity

$$\frac{d}{dt}\left\{\frac{1}{w}\frac{dw}{dt}\right\} = \frac{d^2 \log_e w}{dt^2}$$

symbolizing the specific acceleration of growth, is always negative.[11]

The brightest, though as it happens by no means the most fruitful suggestion in my essays on ageing was that 'heterochronic' transplantation should be used as a means of studying the mechanism of ageing. Heterochronic transplantation—'allochronic' I shall henceforward call it—is the transplantation of tissues between animals of different ages, especially with a view to finding out what happens to the tissues rather than what happens to the animals into which they are transplanted. I had in mind the transplantation of young tissues to older animals and vice versa—a project intrinsically limited by the well-known immunological prohibition upon grafting tissues between animals of different genetic makeup and also, as I now see, the fact that fetal tissues are antigenic to adults of the same genetic makeup; it is not known—though it would be very well worth finding out—if the tissues of senescent animals are antigenic to juveniles. The idea of transplanting juvenile skin to adults was first made by my colleague Rupert Billingham and myself.[12] The idea was that considerations of incompatibility could be circumvented by storing a piece of juvenile skin in the deep freeze until its owner grew up, whereupon the stored skin could be transplanted back to its original owner. There is a photograph of just such a piece of skin on p. 170 of the Ciba monograph on ageing referred to above (7). The use of highly inbred animals makes these experimental tricks unnecessary. The technique of allochronic transplantation has been used to ascertain the relative susceptibilities of juvenile and old skin to chemical carcinogenesis.[13] The most extensive use of allochronic transplantation, however, has been that of P. L. Krohn.[14]

Another essay in this book that deals with evolutionary matters is *The Imperfections of Man*, which was written to correct the belief, at that time quite widespread in medical circles, that natural dispositions are of necessity the best and that 'nature's way' is that which we should at all times attempt to simulate. On the contrary, I pointed out, evolution is very much a fallible, makeshift affair, and loss of fitness in one regard is often the charge for some more-than-compensating gain. In rereading this chapter I noted with amusement that polymorphism—i.e., the stable differentiation of a population into distinct genetic types of which even the least frequent cannot be kept in being merely by the pressure of recurrent mutation—is still a somewhat controversial matter in the sense that there are still those who maintain that it is always genetically enforced, e.g., caused by some such mechanism as a selective advantage of heterozygotes over the corresponding homozygotes, while others adopt the 'neutralist' view that genetic differentiation may arise from as it were casual variations conferring no particular selective advantage or disadvantage.

I had thought myself on pretty safe ground in saying that Man's upright posture 'though a source of moral satisfaction' made him specially vulnerable to lesions of the spinal cord, but much to my surprise this view brought forth a most indignant riposte from a veterinarian. It was clear from the tone of his letter that my proposal that human beings might be the victims of ailments not also suffered by the brute creation was yet another attempt to diminish the standing of the veterinarians—to impede their long struggle for gentility. I sent him a prevaricating but I hope civil answer in which I explained that there was a difference between generalisations that were probably true of the common run of mankind and the special ailments afflicting the mutant horrors whose existence is connived at by the breed societies and show judges.[15] In retrospect, having studies the matter very much more deeply than before, I believe that my contrast between the process of wound healing as it occurs in animals such as rabbits with a mobile integument and as it occurs in human beings proves sound. Upon it is based the compelling necessity for devising some means of making 'allografts' of skin acceptable—i.e., grafts of skin transplanted from one person to another. This is the subject of the last essay in the book, but before turning to it I should mention the other lecture with an evolutionary theme: *Tradition: The Evidence of Biology*.

From the standpoint of the history of ideas this little essay is interesting for two reasons. It was one of the first ventures in a field of speculation that has since become distressingly popular—that which deals with the lessons for human beings of what has been learnt from animal behaviour by the new ethological methods inaugurated by Konrad Lorenz and Nikolaas Tinbergen. As such, it doesn't go very far and perhaps teaches no lesson except the necessity of caution for arguing from animal to man. Fortunately, however, it has all been done much better since.[16] A second reason why this essay may be thought to be of more than ephemeral interest is the discussion of 'exosomatic' or 'exogenetic' evolution; this, too, has been done much better since.[17, 18] Exogenetic evolution, reversible and Lamarckian in style, has often been called 'cultural' evolution, but I think this terminology is injudicious because it might be thought to refer to an evolution of culture instead of an evolution mediated through culture. The notion of this alternative pathway of evolution, peculiar to human beings, was propounded by no less an authority than Thomas Hunt Morgan in a book of lectures entitled *The Scientific Basis of Evolution.*[19]

C. D. Darlington wrote most scathingly of Morgan's idea, saying that it opened the door to Lamarckism and Lysenkoism, a remark which strikes me as being on all fours with blaming Gregor Mendel for the gas chambers. I myself go along with Morgan: there *is* something different about human beings—and biologically speaking this form of evolution is it.

No essay in this book has been farther outdistanced by the march of science than the final essay, *The Uniqueness of the Individual*, which dealt with the vexed and vexing problems of the attempted transplantation of human tissues between one individual and another. Several surgeons have told me that it was this essay that attracted them into the practice of surgery.

The increasing popularity of the reaction against what people have called 'mechanised medicine' provoked by Ivan Ilitch's *Medical Nemesis* prompts me to make a few preliminary remarks about the role of medicine in the treatment of burns. The grave medico-surgical problems raised by the treatment of extensive burns—problems, for which, I argue, skin grafting is the only solution—are in the sense the product of advances in medicine.

In the old days, which weren't so very long ago, a person who suffered a burn extending over 50 per cent. or more of his body surface was not a medical problem: he simply died; but thanks to advances in medicine of a kind we are now widely invited to deplore—I have in mind particularly the institution of the blood transfusion services, the introduction of anti-bacterial drugs and the entire apparatus of intensive care—these victims now live and do give rise to the problems I discuss in my last chapter.

The 'colleagues' I refer to repeatedly in the text—no scientist can have had better—were Dr Rupert E. Billingham, now of the University of Texas, and Dr Leslie Brent, now of St Mary's Hospital Medical School in the University of London. It was they who discovered the causes and consequences of 'graft versus host disease'—the most important new manifestation of the immunological response discovered since my essay was written. In discussing and dismissing the possibility that the acceptance of foreign grafts and therefore the formation of an individual compounded of cells of two different genetic origins might have evil consequences I called attention to the special difficulties that would arise if the host accepted the graft but the graft would not accept the host. 'Graft versus host disease' represents just such a counterreaction of the graft against the host: it arises when mature lymphoid cells from a mouse D are injected into a mouse of different genetic makeup R which is for any reason—whether natural or artificial—incapable of defending itself against D cells. Under these circumstances D lymphocytes attack the tissues of R and may kill R: they are doing just what they would do if R grafts from R had been transplanted onto D, but in this situation D's lymphocytes are playing an away match. The study of the various manifestations of 'graft versus host disease' and of their relevance to human disease has been one of the most rewarding and fruitful developments of modern transplantation biology.

The three of us—my two colleagues and I—worked together very closely in demonstrating the phenomenon of immunological tolerance, described in the text in terms that by modern standards seem painfully simplistic. The modern clinical use of transplants did not grow directly out of this demonstration of 'actively acquired tolerance'; no, the importance of the discovery of tolerance was essentially a moral one: it showed for the first time that the barriers that normally prohibit the transplantation of tissues from one individual to another *can* be broken down—there was indeed no natural embargo upon the act of transplantation. I remember very well how this discovery raised everybody's spirits and sent them back to their laboratories or wards with renewed faith that the transplantation problem would eventually be overcome. In retrospect I think the greatest contribution of transplantation to biological science and to human medicine is that which grew out of the work of G. D. Snell and his colleagues and Peter Gorer and his colleagues on the genetic basis of tissue compatibility in mice. The great importance of this work has not merely been to facilitate the act of transplantation and to put it on a footing analogous to the blood transfusion but rather to uncover the existence in human beings of an entirely new system of polymorphism—of stable differentiation into different genetic types which can now be recognized by typing procedures analogous and distantly akin to those that reveal conventional blood groups.

Work on mice provided the techniques and the conceptual background for the typing of human beings and this in turn has led to the recognition of the genetic basis of differences in susceptibility to multiple sclerosis, ankylosing spondylitis and the juvenile (insulin-dependent) form of diabetes. This great accession of understanding was not easily won: tens of thousands of man hours were spent upon experiments on tumour transplantation in mice which made no kind of sense until men such as C. C. Little, and G. D. Snell and their colleagues took the matter in hand and worked out the genetical theory of transplantation.

This story has a profound moral: how could such a discovery as that of the genetic background of the above-mentioned diseases have been premeditated? What advice should have been given to some young grant applicant who sought to uncover the genetics of juvenile diabetes and multiple sclerosis? Why is it that after the event one can see—what is quite absurd—that he should have been told to go off and study the genetics of tumour transplantation in mice and work out reliable serological methods of allocating a mouse to one type rather than another; this propadeutic exercise complete, the investigator should then have been advised to apply these techniques to human beings, and thus uncover a system of polymorphism in human beings analogous to that which governs the outcome of tissue transplantation in mice. Having done this he could turn to finding out if one tissue transplantation type or another were disproportionately frequent or infrequent in the victims of some particular disease. This scenario is of course a *reductio ad absurdum* of the notion that scientific discovery can be premeditated, as if by the application of 'the scientific method'.

I am indebted to Sir John McMichael, FRS, for an even better example. With the coming of anaesthesia and aseptic surgery the therapeutic powers of surgery, particularly in the domain of gastroenterology, increased by leaps and bounds, but of course the surgeon was handicapped by not knowing in advance what was amiss inside his patients. What therefore could be a more important research project than a proposal to discover a method for making human flesh transparent? Suppose now that a project grant review system similar to that which prevails today were in force at the turn of the century. It rates no great feat of imagination to picture in one's mind the regretful, almost pitying shakings of the head that would accompany the verdict that such a proposal was idiotic in the extreme. Imagine, too, the meeting of a peer review committee as they judged their colleague to be in need of psychotherapy. And, of course, put as I have put it, the project is indeed impractical; but the fact of the matter is that a method of making human flesh transparent was discovered round about the turn of the century by Roentgen and others in the course of the ordinary process of scientific discovery.

The moral is that although the 'ordinary processes of scientific discovery' are culpably wasteful and inefficient, and are looked upon with contemptuous disfavour by people who have never made a discovery of any importance in their lives, we cannot yet do better; *scientific discovery cannot be premeditated.*

I can see now that I was unduly confident in saying that the antigens which arouse immunity against 'allografts' (as homografts are now called) enter the lymphatics and are then wafted into the regional lymph nodes; it now seems to me more likely that the process of sensitization comes about when the graft is 'visited' by circulating blood lymphocytes, which are thereupon activated by the foreign antigen and clamber into the original lymphatics and so find their way into the regional nodes.

Among the passages omitted in accordance with the principle I enunciate at the end of this Introduction were a number of quite unduly confident pronouncements about the nature of antigens, i.e., the stimuli that arouse transplantation rejection; I am consoled, however, by the thought that even today it is not possible to make a definitive statement about the chemical nature of transplantation antigens.

In general, it would not now be doubted that transplantation biology—the first branch of experimental medicine to encourage the recruitment into it of zoologists and general biologists—has made some profound contributions to immunology in particular and to experimental medicine in general. The discovery of the nature and functions of the thymus gland by R. A. Good and J. F. A. P. Miller grew directly out of the study of transplantation; so also did the discovery that the immunological response has two quite distinct arms: there is a *humoral* immunity transacted by one class of lymphocytes, and a cell mediated immunity by a second class of lymphocytes which depend for their maturation upon the influence of the thymus gland. The former, in particular, was a remarkable achievement which has not received the attention it deserves from biologists generally: the thymus is a large and important-looking organ, especially prominent in young animals, including human beings. Of no other organ of comparable size can it be said that the discovery of its nature and functions had to await the middle of the twentieth century.

Beyond this fine achievement, transplantation biology may also be credited with discovering the existence of the new modality of the immune response I referred to above, the graft versus host reaction: in addition, as I said, transplantation biology led to the discovery of the existence in human beings of a system of genetic polymorphism quite distinct from and at least as important as that which is associated with the blood group of antigens.

In spite of all the errors of fact or judgement which I have been able to discern in this text—only a fraction, I gravely fear, of those that might be discovered by a critic determined to find fault—I still have a lively affection for this book simply because it was full of ideas, not all of which were mistaken.

A book such as this written today by one such as I then was would disclose an entirely different spectrum of preoccupations and burning interests, certainly not less fascinating but probably not more: it is just these 'dissolves' and changes of intellectual scenery that make the history of ideas so fascinating a subject, and this is my only real excuse for contriving to give this book a second innings.

Apart from this new Introduction, written especially for the Dover edition, this book is reproduced as I wrote it except for the omission of passages that were plainly erroneous or interpretations that subsequent events have shown to be gravely at fault. I am not so far wedded to the idea of antiquarian authenticity as to condone the perpetuation of ideas that are frankly wrong.

P. B. MEDAWAR

London, 1981

REFERENCES

1. Thompson, D'Arcy Wentworth. *On Growth and Form.* Cambridge University Press, 1917.
2. Wright, Sewall. *Evolution and the Genetics of Populations*, vols. 1-4. Chicago University Press, 1968-78.
3. Zhores, A. Medvedev. *The Rise and Fall of T. D. Lysenko.* Trans. by I. Michael Lerner. Columbia University Press, New York, 1969.
4. Williams, G. C. 'Pleiotropy, Natural Selection, and the Evolution of Senescence'. *Evolution*, **11** (1957), pp. 398–411.
5. Kirkwood, T. B. L. 'The Evolution of Ageing'. *Nature.* **270** (1977), pp. 301–304; 'Senescence and the Selfish Gene', *New Scientist*, 29th March (1979), pp. 1040–1041; Kirkwood, T. B. L. and Holliday, R. *Proc. Roy. Soc. B*, **25** (1980), pp. 531–545.
6. Pyke, D. *Diabetologia*, **17** (1979), pp.333–343.
7. *Ciba Foundation Coloquia on Ageing Vol.* 1: *General Aspects.* Churchill, London, 1955.
8. Coale, A. J. 'Age Composition in the Absence of Mortality and in Other Circumstances'. *Demography*, **10** (1973), pp. 532–542.
9. Hayflick, L. 'The Limited *in vitro* Lifetime of Human Diploid Cell Strains'. *Experimental Cell Research*, **37** (1965), pp. 614–636; Martin, G. M., Sprague, C. A. and Epstein, C. J. 'Replicative Lifespan of Cultivated Human Cells: Effects of Donor's Age, Tissue and Genotype'. *Laboratory Investigation*, **1** (1970), pp. 86–92.

10. Lansing, A. I. 'A Cumulative Factor in the Ageing Process'. *Anatomical Record*, **96** (1946), pp. 539–540.
11. Medawar, P. B. 'Size, Shape and Age'. In *Essays on Growth and Form presented to D'Arcy Wentworth Thompson*, Ed. by W. E. le Gros Clark and P. B. Medawar. Clarendon Press, Oxford, 1945.
12. Billingham, R. E. and Medawar, P. B. 'The Freezing, Drying and Storage of Mammalian Skin'. *J. exp. Biol.*, **29** (1952), pp. 454–468.
13. Ebbeson, P. 'Ageing Increases Susceptibility of Mouse Skin to DMBA Carcinogenesis Independent of General Immune Status'. *Science*, **183** (1974), No. 4121, pp. 217–218.
14. Krohn, P. L. 'Heterochronic Transplantation in the Study of Aging'. *Proc. Roy. Soc. B.*, **157** (1962), pp. 128–147.
15. Medawar, P. B. 35th Stephen Paget Memorial Lecture: 'Animal Experimentation in a Large Research Institute'. *Conquest*, LV (1967). No. 158.
16. See *Growing Points in Ethology*, especially the article 'Does Ethology Throw Any Light on Human Behaviour?', pp. 497–506. Ed. by P. G. Bateson and R. A. Hinde, Cambridge University Press, 1976.
17. Medawar, P. B. and Medawar, J. S. See *The Life Science*. Harper and Row, New York, 1977.
18. 'Technology and Evolution'. Pp. 105–115 in *The Frontiers of Knowledge*. Doubleday, New York, 1975.
19. See Allen, G. E. *Thomas Hunt Morgan: A Scientific Biography*. Princeton University Press, 1978.

1

Old Age and Natural Death

The problems of old age and natural death are hardly yet acknowledged to be within the province of genuine scientific enquiry. This does not mean that biologists are ignorant of the fact that such problems exist, nor that natural death is altogether insusceptible of scientific treatment. It simply means that no such treatment has been given it yet.

This neglect is partly the outcome of a certain quickening in the tempo of biological research. The biologist of to-day is a busy man: he has no time for anecdotes about the age of tortoises, and wants more evidence than Metchnikoff had power to give him before he takes steps to modify the flora of his bowel. Yet nearly all the great theorists of the last century were fond of teasing themselves with speculations about death. '*Qu'est ce que la vie?*' Claude Bernard[1] asked himself: '*La vie, c'est la mort.*' Life is combustion, and combustion death. '*La vie est un minotaure, elle dévore l'organisme.*' This is only one of alternative views on the nature of natural death. The distinction of first suggesting that natural death might be an epiphenomenon of life, rather than something of the very nature

[1] *Définition de la Vie* (1875); one of the essays reprinted in *La Science Expérimentale*, 7th ed., Paris, 1925. Also Bernard quoting Buffon.

of the act of living, is shared unequally between August Weismann[1] and Alfred Russell Wallace. (Wallace's views are known to us only from a casual letter preserved by Edward Poulton.[2] They are about the same as Weismann's, though less confidently and much less lengthily expressed.) But before I try to give an account of Weismann's views, we must have a few definitions; for the trouble with 'natural death' is not that it lacks a meaning, but that it has the embarrassment of two or three. By 'accidental death', then, or simply 'death', is meant death from any cause whatsoever. 'Natural death' is that sort of death by accident to which the age-specific decline of our faculties, *senescence*, has made a certain contribution, however small. The contribution grows larger as we grow older: what lays a young man up may lay his senior out: but it always falls short of unity, for no one dies merely of the weight of years. Ludwig Aschoff, the greatest clinical pathologist of the last generation,[3] looked back upon his life for evidence of such a case. He once thought he had found it in a colleague ninety-four years old, whose life seemed merely to fade away; but autopsy showed a lobar pneumonia of four days' standing.

We shall be obliged to use the term 'natural death' in the rather wide sense of the foregoing definition. Popular usage quite rightly fines it down to forms of death to which senescence has made a pretty big contribution, for it seems absurd to say that a man of forty could die in part of old age. At all events, what we are to discuss is not the event, death, but the *process* of senescence. (The definition of death itself, in the most familiar of its several meanings, can be valid only with

[1] *The Duration of Life* (pp. 1-66) and *Life and Death* (pp. 111-61); essays reprinted in *Weismann on Heredity*, ed. E. B. Poulton, S. Schönland, and A. E. Shipley; 2nd ed., Oxford, 1891.

[2] *The Duration of Life*, pp. 23-4.

[3] Cf. *Lancet*, **235**, p. 87, 1938.

reference to some stated 'level' of biological organization. A society will die before its individual members, an individual before his cells, and a cell before its ferments have stopped working. But legally, I suppose, a man is dead when he has undergone irreversible changes of a type that make it impossible for him to seek to litigate.)

Weismann believed that natural death had evolved under a Darwinian regimen of natural selection. The 'utility of death' he says, is this. 'Death takes place because a worn-out tissue cannot for ever renew itself. . . . Worn-out individuals are not only valueless to the species, but they are even harmful, for they take the place of those which are sound.' It follows that 'by the operation of natural selection, the life of a theoretically immortal individual would be shortened by the amount which was useless to the species'.[1] In this short passage, Weismann canters twice round the perimeter of a vicious circle. By assuming that the elders of his race are decrepit and worn out, he assumes all but a fraction of what he has set himself to prove. Nor can these dotard animals 'take the place of those which are sound' if natural selection is working, as he tells us, in just the opposite sense. It is curious that Metalnikov in his comparatively recent *La Lutte contre la Mort* (1937) should give these fallacies a seventy-five-year run by twice repeating them with approval word for word. The problem is, *why* are the older animals decrepit and worn out? And for this Weismann had no sufficient answer. It must be obvious that, senescence apart, old animals have the advantage of young. For one thing, they are wiser. The Eldest Oyster, we remember, lived where his juniors perished. They are wiser, too, in their experience of infection, for an animal which has survived a first infection is better equipped to deal with it a second time. In the majority of animals 'immunological wisdom' may be a better bargain

[1] *The Duration of Life*, p. 24.

than anything they may have by way of mind. We are always inclined to over-estimate the value of mental wisdom, though no one, I suppose, has the temerity to doubt that the giraffe owes more to his long neck than to the organ poised on top of it; and the logic of brute fact tells us that the extinct reptile, *Diplodocus*, which had a brain in the pelvic region as well as up in front, drew little advantage from his power to reason not merely *a priori*. No: what kills the old animal is not in the first place decrepitude, but something which has the dimensions of the product of time by luck.

Weismann had a theory not merely of the evolution of death in animal populations but also of the mechanism of senescence in the individual. He believed that a limit to life was set by an inherent limitation in the power of germ cells to divide. 'We do not know', he says, 'why a cell must divide 10,000 or 100,000 times and then suddenly stop,'[1] as he thought it did. As a matter of fact, we now know that no such inherent limitation exists; but Weismann's theory—if we disregard the fact that the progeny of a cell which divided only 10,000 times would fill the utter limits of known space—shows that he had not appreciated the 'asymptotic' character of the process of age-decline. He had no grasp of the *process* of ageing. We don't grow old suddenly, and the cells within us do not suddenly stop dividing. Those that do stop come to rest in a decent orderly fashion. Charles Minot[2] was the first to make this clear. He took over Weismann's idea that death had evolved by natural selection, and turned his mind to ageing in the individual alone. His views were original and still are theoretically important, so they deserve a fuller treatment than they commonly get.

[1] *Ibid.*, p. 22.

[2] *The Problem of Age, Growth, and Death*, London, 1908; a series of lectures first published in *Popular Science Monthly*.

Minot used growth as a measure of vitality; not the mere rate of growth, but the *specific* rate, which gives us a measure of the capacity of living tissue to reproduce itself at the rate at which it was formed. It is simply the rate of growth at any chosen time divided by the size at that same time—in other words, by the material theoretically available for further growing. It cannot be denied that the specific growth rate *is* a measure of vitality, though not perhaps so complete a measure as Minot in his time believed. Minot found that the power of tissue to reproduce itself at the rate at which it was formed fell off through life from earliest childhood onwards. He found that the decline was faster in children than in their elders, and, indeed, that it fell off more and more slowly as life went on. The inferences he drew were these. There is no period of increasing vitality leading to the mature state and thereafter to the senile; the process of ageing goes on continuously throughout life. And ageing is faster in young animals than their elders—'a strange, paradoxical statement'. 'Our notion that man passes through a period of development and a period of decline is misleading . . . in reality we begin with a period of extremely rapid decline, and then end life with a decline which is very slow and very slight.'

This is a good moment to ask what the life insurance companies have to say about these problems. Their evidence is at first sight very helpful. Look at the curve from which the actuary computes the force of mortality at various ages—the curve which defines, for each age of life, the numbers still living of a certain initial number born alive. From the twelfth or fifteenth year onwards in human life, the curve is smooth; there is no break or discontinuity, no hint at all that at such an age the prime of life has ended and old age begins. Nor is this generalization false for animals other than man. 'Life tables' for them are pitifully meagre; but Leslie and Ranson made one lately for the laboratory vole,[1] and here too we find the same smooth passage to extinction. 'Voles drop off at all times of life,' said Elton,[2] speaking of this evidence,

> 'though not at the same rates. And these are not "ecological" deaths; few of them probably are "parasitological" deaths. We hardly know what process is at work, and for want of a better term we may call it "wear and tear". This has the suggestion of an internal breakdown in the physiological organization. *We might almost say that the process of senescence begins at birth.*'

This final inference, which I have italicized, is by no means immediate. The actuary's life table is not a mapping of the course of individual life: it is founded on the *distribution* through life of the ages at which people die. It thus relates to no event in life save one, its end. Even if the sudden flowering of an evil gene caused voles to age and die within a day, the ages of their deaths might well be so pieced out among the population as to yield just that smooth, continuous curve the actuary maps for us. If, however, the population is reasonably uniform, then the life table (or rather, the force of mortality computed from it) does indeed give us what may be called a 'statistical picture' of the course of ageing. For we may define 'senescence' as that which predisposes the individual to death from accidental causes of random incidence; and it follows that the frequency distribution of the ages of death gives us a

[1] P. H. Leslie and R. M. Ranson, *Journal of Animal Ecology*, **9**, p. 27, 1940. For life tables for invertebrate animals, cf. A. J. Lotka, *The Elements of Physical Biology*, Baltimore, 1925; W. H. Dowdeswell, R. A. Fisher, and E. B. Ford, *Annals of Eugenics*, **10**, p. 123, 1940; C. H. N. Jackson, *ibid.*, p. 332. Jackson finds that the life table of tsetse flies is biased, during the rainy season only, by an element contributed by senescence. [For further evidence see *Principles of Animal Ecology*, by W. C. Allee, O. Park, A. E. Emerson, T. Park and K. P. Schmidt (Philadelphia, 1949); E. S. Deevey, *Quart. Rev. Biol.*, **22**, p. 283, 1947; *The Natural Regulation of Animal Numbers*, by David Lack (Oxford, 1954); *The Biology of Senescence*, by Alex Comfort (London, 1956).]

[2] C. S. Elton, *Voles, Mice and Lemmings*, pp. 202-5, Oxford, 1942.

statistical picture of the magnitude of this predisposition. Many sciences use a picture of this sort, and some use no other; the problems it raises are interesting, but not at the moment relevant.

Minot wanted to bring not merely size but shape as well within the ambit of his laws; but complained, as many have done since, that 'we do not possess any method of measuring differentiation which enables us to state it numerically'. Such attempts as have been made to do so support his theory; for example, the rate of change of shape of the human being falls off progressively through life.[1] But we do know that Minot's laws are by no means commonly true of faculties other than those which turn upon the pattern and the rate of growing. The sort of sensory, motor and 'mental' tests that are used to measure physical and intellectual prowess usually give their best values in the neighbourhood of the age of twenty-five, or later. Usually, but not always: it is around the age of ten that hearing of very high-pitched sounds is most acute.[2] Information of this sort is intrinsically important, for it does something to confirm a theorem of wide significance which many clinicians have long taken for granted—that the time of onset and rate of ageing of the faculties and organs may vary independently within fairly wide limits. Other evidence tells against it. One of the most useful lessons to be learnt from the natural historian's studies of animal longevity[3] is that the life span varies greatly in length between quite closely related types of organism. What can this mean, if not that the ageing process in the individual as a whole is geared by one or two limiting or 'master' factors?

Minot's special theory of the ageing process is just as unusual as are his general laws, for he believed that cellular

[1] P. B. Medawar, *Proceedings of the Royal Society*. Series B, **132**, p. 133, 1944.

[2] Y. Koga and G. M. Morant, *Biometrika*, **15**, p. 348, 1924. Cf. the data summarized by V. Korenchevsky: *Annals of Eugenics*, **11**, p. 314, 1942.

[3] The most important of these are by S. S. Flower. See note 2, p. 32.

differentiation is the cause of the progressive fall away of growth potential. Cellular differentiation—the degree of muscliness of a muscle fibre, for example—has never been measured, but Minot guessed that if such a measurement were to be made, the curve of increasing differentiation would be found to be the exact complement of that which plots the declining energies of growth. To put it in another way: that which we call 'development' when looked at from the birth end of life becomes senescence when looked at from its close. It is an attractive idea, but such little evidence as we have speaks against it. The tissue cultivator, who grows cells in blood and tissue media outside the body, finds that 'old' cells have just as high a *capacity* for growth as young ones. They simply take a longer time to set about it.[1] It is perfectly true that some very highly differentiated cells, like those of nerve and muscle, lose their power to multiply by fission. But that is more of a mechanical accident than a slur upon their vitality; after all, a nerve cell may be some yards long. Neither adult nerve nor adult muscle has lost the power to *grow*, and if a muscle or nerve fibre is cut into two, healing and replacement will start up from one end or the other. But whatever the rights and wrongs of Minot's special theory, he has left us with two ideas which any future theory of the ageing process must analyse and suitably explain: the first is the continuity of the ageing process, the second its great span in time.

Some mention must now be made of the celebrated and widely misinterpreted views of Elie Metchnikoff on ageing.[2] Metchnikoff believed that much of what in ageing seems to us to be very 'natural' is in fact abnormal. How much of ageing he held to be so is far from clear, though he seemed to think, as Buffon did, and later Flourens, that an animal's total span

[1] Cf. the evidence summarized by P. B. Medawar, *Proceedings of the Royal Society of Medicine*, **35**, p. 590, 1942.

[2] *The Prolongation of Life* (trans. P. Chalmers Mitchell), London, 1910.

of life should be between five and seven times the period that passes between birth and the onset of sexual maturity. Self-intoxication by the products of bacterial decomposition in the large intestine was chiefly to blame for the pathological changes of senescence. The theory has a homely origin. The mammals, Metchnikoff argued, do not void their faeces on the run, and yet are exposed to countless dangers by doing so when standing still. In order to choose the most appropriate time for defaecation, mammals must therefore have large intestines in which to store their faeces.[1] Bacteria flourish in the store-house so provided, and the absorption of their evil humours brings about a state that ranges from the malaise of constipation to the chronic and cumulative toxaemia of pathological senility. Cells intoxicated beyond redemption are attacked and eaten up by the phagocytic cells which, conveniently enough, Metchnikoff himself had earlier discovered.

Most laymen are convinced that there is something in this theory, and it has not lacked zoological champions of the greatest eminence. 'Certain it is,' said MacBride[2] some twenty years later, in the course of a violent attack on mathematical biology, 'certain it is that in human beings, when the toxins produced by proteolytic enzymes are got rid of, many of the signs of old age may disappear.' But a biologist can pick holes in each single theorem. Some mammals do defaecate while running. The malaise of constipation is at once relieved by bowel movement, and fishermen who habitually defaecate at ten-day intervals are not the debile wrecks that Metchnikoff's theory would have us think them. The large intestine, too, is

[1] It is a popular fallacy that faeces await evacuation in the rectum. This is so only in cases of chronic constipation. Cf. Sir A. Hurst, *Proceedings of the Royal Society of Medicine*, **36**, p. 639, 1943.

[2] In the discussion of G. P. Bidder's Linnean Society lecture on ageing (note 2, p. 31). MacBride had been particularly upset by Karl Pearson's statement that mental deterioration in man began at the age of twenty-seven.

no mere dustbin. Herbivorous animals get some of their food from the action of cellulose-splitting bacteria within it. The bacteria may, moreover, synthesize vitamins, which are absorbed directly or may be recovered by eating the droppings themselves—a slap in the eye for Metchnikoff's theory. The theory is dead, and nothing is to be gained by propping it up into a sitting position.*

In the first twenty years of this century, there began to accumulate new empirical evidence concerning the 'immortality' of the ordinary non-reproductive cells of the body—more exactly, the immortality of the cell-lineages to which, by successive acts of fission, such cells may be ancestral. Leo Loeb and later, more clearly, Jensen showed that several tumours will grow indefinitely if handed on by grafting from one animal to another.[1] It used to be possible to buy from the laboratories of the Imperial Cancer Research Fund a rat bearing Jensen's rat sarcoma. Its cells are lineal descendants of those which Jensen first transplanted some forty years ago. The technique of growing cells outside the body proved as much for the cells of normal tissue. A strain of connective-tissue cells was started by Carrel and Burrows in 1912.[2] The first year's growth was not enough to demonstrate the perpetuity of the cell lineage. We are 'not justified', said Ross Harrison in 1913, 'in referring to the cells as potentially immortal . . . until we are able to keep the cellular elements alive in cultures for a period exceeding the duration of life of the organism from which they are taken. There is at present no reason to suppose that this cannot be

[1] A clear elementary account of this early work is to be found in W. H. Woglom, *Fifth Scientific Report of the Imperial Cancer Research Fund*, p. 43, London, 1912.

[2] There are quite a number of popular accounts of this work, e.g. in A. Carrel, *Man the Unknown*, New York, 1935; L. du Nöuy, *Biological Time*, London, 1936.

* [A comparatively recent paper on Metchnikoff's theory is that by S. Orla-Jensen, E. Olsen and T. Geill, *Journal of Gerontology*, **4**, p. 5, 1948.]

done, but it simply has not been done as yet.' In due course it was done, and the strain was with us until 1939. Tissue-culture has other evidence to offer us of death. We are told that one of the last experiments of Thomas Strangeways was to cultivate the connective-tissue cells surviving in a sausage—as neat a demonstration as one could wish of the tenacity of the *vita propria* and the half-truth that is legal death. So let us submit yet another zoological simile of common speech to the censorship of our new wisdom. The earth stirs over Mendel's grave when we say that two people are *as like as two peas*. Many fish, moreover, never drink. 'As dead as mutton' is likewise superannuated by the march of time; and those whose most pressing fear it is that they will be lowered living into their graves can have their doubts resolved: they will be.

(The so-called 'immortality' of the Protozoa is like that of the tissues: not an immortality of cells but an indeterminateness of cell lineages. Obviously the cell lineages of protozoa are in some cases immortal or indeterminate, for otherwise they could hardly be with us to-day. But does this immortality depend upon the performance of an occasional act of nuclear reconstitution, or can protozoa thrive for ever by the mere act of dividing asexually into two? The matter has long been controversial.[1] Some of the early investigators believed that, in default of such 'rejuvenation', a protozoan lineage must undergo a microcosmic cycle of growth, maturity, decay and death, exactly like the cell population of higher organisms. Others believed that vegetative fission would suffice. When it came to be known that the former opinion was founded at least in part on the use of faulty techniques of cultivation, the latter dispossessed it. But Jennings[2] is now inclined to doubt whether asexual fission is in itself enough, and the more recent genetic

[1] Cf. H. S. Jennings, *Problems of Ageing*, ed. E. V. Cowdry, 2nd ed., pp. 24-46, Baltimore, 1942.

[2] *Journal of Experimental Zoology*, **99**, p. 15, 1945.

evidence suggests that some sort of nuclear rehabilitation is from time to time required. Ordinary asexual fission is, from the mechanics of the process, a very exact division of the parent organism into equal parts. The genetical sins of the parents—the lethal or unwholesome mutant genes—are thus allotted to their progeny with biblical justice and more than biblical precision. The nuclear reconstitution spoken of above is, in effect, a device by which such genes may be eliminated from the stock. The organisms which inherit them die soon, or fail to reproduce; the others, often a minority, carry on.)[1]

With such new facts as these at his disposal, and others of great value added by himself, Raymond Pearl[2] made the next important attack on death in 1922. Pearl himself showed that an animal's span of life was governed by inherited factors and was within certain limits subject to experimental modification. The total span of life may be increased not by adding a few extra years to its latter end nor, if it comes to that, by intercalating new life at any intermediate period, but rather by stretching out the whole life span symmetrically, as if the seven ages of man were marked out on a piece of rubber and then stretched. The length of life may thus be treated as a function of the *rate* of living. One simple way of lowering the rate of living—an ingredient of many a centenarian's recipe for long life—is to withhold with known precision the sort of food that is used for the supply of energy: a restriction of calories, as we say, rather than a systematic *mal*nutrition. McCay and his colleagues[3] have shown that by such means the life span of rats may be greatly lengthened. The same is true of flatworms, as

[1] Cf. B. F. Pierson, *Biological Bulletin*, **74**, p. 235, 1938; T. M. Sonneborn, *ibid.*, p. 76. I am obliged to Professor J. B. S. Haldane for pointing out the significance of their evidence. [For more recent evidence on senescence in protozoa, see Alex Comfort, *The Biology of Senescence*, London, 1956.]

[2] *The Biology of Death* (Lowell Lectures), Philadelphia, 1922; *The Rate of Living*, London, 1928.

[3] Cf. C. M. McCay, pp. 680-720, in *Problems of Ageing* (note 1, p. 27).

Child told us;[1] of certain sea-squirts, and of the aberrant, worm like creatures known as Nemertines.[2] These latter have the advantage of the rat, for if deprived of food they react by growing smaller, thus literally retreating into second childhood. They do not quite exactly retrace their steps, 'advancing backwards' (as was said of a recent famous military campaign) along the path they followed in development; but in a sense they cheat Time. The fact that starved rats outlive those which habitually eat sufficient is often used as evidence of the relativity of biological time; but in reality, it is evidence less of the tortuous mysteries of time and space than of the virtues of sobriety and moderation.

In the extreme case, when life is held altogether in abeyance, we may properly speak of immortality. Freeze a tissue such as mammalian skin to the temperature of liquid air (something less cold will do) and the resumption of life will then await the convenience of the experimenter.[3] The idea is an old one. Until he tried to freeze two carp, John Hunter—[4]

> 'imagined that it might be possible to prolong life to any period by freezing a person. . . . I thought that if a man would give up the last ten years of his life to this kind of alternate oblivion and action, it might be prolonged to a thousand years; and by getting himself thawed every hundred years, he might learn what had happened during his frozen condition. Like other schemers, I thought I should make my fortune by it; but this experiment undeceived me.'

[1] C. M. Child, *Senescence and Rejuvenescence*, Chicago, 1915.

[2] See J. Needham, *Biochemistry and Morphogenesis*, pp. 524-9, Cambridge, 1942.

[3] Cf. R. Briggs and L. Jund, *Anatomical Record*, **89**, p. 75, 1944; J. P. Webster, *Annals of Surgery*, **120**, p. 431, 1944. The author has often confirmed their observations. [See R. E. Billingham and P. B. Medawar, *Journal of Experimental Biology*, **29**, p. 454, 1952.]

[4] J. Hunter, *Of the Heat of Animals*, in *The Works of John Hunter, F.R.S.*, ed. J. F. Palmer, Vol. 1, p. 284. The phenomenon which Hunter unluckily failed to demonstrate has been called 'anabiosis'.

These particular carp died, though latter-day experimenters have been more lucky.[1]

Raymond Pearl agreed with Weismann that in some manner or other natural death had evolved, but that it evolved under the auspices of natural selection he irritably denied. ('Probably no more perverse extension of the theory than this was ever made.') Yet for so brilliant a man, Pearl's own theory of the mechanism of ageing in the individual is curiously inadequate. 'Specialization of structure and function necessarily makes the several parts of the body mutually dependent for their life upon each other. If one organ or group, for any accidental reason, begins to function abnormally and finally breaks down, the balance of the whole is upset and death eventually follows.' But is not this a description of the 'proximate cause' of almost any form of death? Something gives way, no doubt: one man will be as old as his arteries, another as his liver. But gross abnormality apart, why should any organ break down? Apparently because of the wear and tear of merely working, and Pearl tells us that 'those organ systems that have evolved farthest away from the original primitive conditions . . . wear longest under the strain of functioning'. It is only towards the end of his book that Pearl puts forward his theory in this relatively specific form. Earlier—and see how much more easily he breathes the air of amorphous generalization—he tells us that the somatic death of higher organisms 'is simply the price they pay for the privilege of enjoying those higher specializations of structure and function which have been added on as a sideline to the main business of living things, which is to pass on in unbroken continuity the never-dimmed fire of life itself'.

[1] E.g. N. A. Borodin, *Zoologische Jahrbuch*, **53**, p. 313, 1934. [To-day, thanks to the work of R. Andjus of the University of Belgrade, even mammals can survive being frozen: the subject of 'hypothermia' and of tissue storage by freezing has been admirably reviewed by R. E. Billingham in *New Biology*, **18**, p. 72 (London: Penguin Books, 1955). See also A. U. Smith, J. E. Lovelock and A. S. Parkes, *Nature*, **173**, p. 1136, 1954.]

A stirring thought; but Johannes Müller had said as much some eighty years beforehand[1] and with proper scientific caution had remarked: 'This has the appearance of explaining the phenomena, but is in reality a mere statement of their connection, and it is not even certain that as such it is correct.'

Let us now turn to one last famous speculation on the problem of natural death. Minot, we saw, left us with the capacity for growth as an upside-down measure of the rate of ageing. Suppose an animal increased in size indefinitely: would it die a natural death? Hardly, if so important a function as growth were left undimmed by age. But before hearing Bidder's answer,[2] the question can be put a little more exactly. The distinction is not between animals which continue to grow and animals which stop growing but between animals without and with a *limit* to their size. How the limit is approached is neither here nor there. It may be approached asymptotically, as in mathematical theory, or finally—to a maximum—as for all intents and purposes it is in fact. According to Bidder, fish grow without limit and never undergo senescence nor suffer natural death. Indeed, he does not 'remember any evidence of a marine animal dying a natural death'. Now a mechanical limit is set to the size of animals on land, as Galileo and many others since have taught us; and according to Bidder this limit is set, or has come to be set, by an intrinsic limitation of the power of growth, with senescence as its outcome. 'Did old age and death only become the necessary fate for plants and animals when they left the swamps, claimed the land, and attempted swiftness and tallness in a medium $\frac{1}{800}$ of their specific gravity?' Bidder believes that this is so, if the quite

J. Müller, *Elements of Physiology* (trans. W. Baly), Vol. 1, pp. 35-6 (and cf. Vol. 2, p. 1660), London, 1840-2.

[2] G. P. Bidder, *Proceedings of the Linnean Society*, p. 17, 1932, *British Medical Journal*, **2**, p. 583, 1932.

special category of 'parental' death, like that suffered by the male salmon, is left out of count.

We will skip blindfold over the causal nexus that relates the limitation of growth to the degenerative changes of old age, and ask ourselves if Bidder's main thesis, that marine animals do not die natural deaths, is in fact true. It is a 'highly debatable problem'—that is to say, one with so little evidence to its credit that no debate is in reality worth while. We have, it appears, little to say about the death of fish that Ray Lankester did not say in his Prize Essay on longevity some eighty years ago:[1] 'they are not known to get feeble as they grow old, and many are known *not* to get feebler'. 'Real evidence is practically non-existent,' said Major Flower,[2] though he could tell us that 'under favourable circumstances some fresh-water fishes may live for half a century'. The fact of the matter is that the energy that might have been devoted to a theoretically straightforward solution of the problem has very often been dissipated in digging up anecdotes about longevity from obsolete works of natural history. Nor has the research been theoretically prudent, for often no distinction has been made (though Lankester insisted on it) between the mean expectation of life and the total life span. It proves that we cannot accept the claims of most of the famous human more-than-centenarians, so what faith are we to have in the pedigrees of tortoises and carp? No one has yet made a systematic study of whether even mammals *in their natural habitat* do indeed live long enough to

[1] E. Ray Lankester, *On Comparative Longevity on Man and the Lower Animals*, London, 1870.

[2] See the series of articles in the *Proceedings of the Zoological Society* (latterly *Series A*), 1925, 1931, 1935-8. [Flower's last paper, published posthumously, on the alleged longevity of elephants, should on no account be missed: see the *Proceedings of the Zoological Society*, **117**, p. 680, 1947. The old original Jumbo ('Old Jumbo carried generations of London children round the zoo in Regent's Park') died at 24, Alice at 50, and Napoleon's Elephant at about 53. Flower dissects the legends of their longevity with admirable skill.]

reach a moderate though certifiable degree of senility. As a matter of fact, the contribution that senescence makes to accidental death can be deduced with reasonable accuracy from the mathematical character of the actuary's life table. For if the 'force of mortality' were constant and independent of age; if, that is to say, the chances of dying were the same in the age interval 100-101 years as in the interval 10-11 years; then the curve defined by the life table would be of the familiar die-away type that describes, for example, the loss of heat from a cooling body. But no life table has yet been made for a mammalian species in the wild. All that can be said so far, in the spirit of Lankester's generalization, is that *some* mammals do *not* appear to live that long. Hinton's studies[1] on fossil and recent voles of the genus *Arvicola* showed that 'not only are the molars still in vigorous growth, but the epiphyses of the limb bones are still unfused with their shafts. Apparently, that is as far as actual observation goes, voles of this genus are animals that never stop growing and never grow old. But no doubt, if one could keep the vole alive in natural conditions, but secure from the fatal stroke of accident, a time would come when . . . the animal would become senile and die in the normal manner.' Burt's study[2] of mice of the genus *Peromyscus* led to a similar conclusion; but there, so far as I know, the matter stands. The difficulties of constructing life tables for animals in the wild are technically formidable, but they must be solved.*

From the standpoint of evolutionary biology an animal's expectation of life in its natural surroundings is much more significant than the degree of decrepitude to which it may be nursed in laboratory or zoo. It is a fair guess that much of what

[1] M. A. C. Hinton, *Monograph of the Voles and Lemmings*, Vol. 1, p. 48, British Museum, London, 1926.

[2] W. H. Burt, No. 48 in *Miscellaneous Publications of the Michigan University Museum of Zoology*, May, 1940. I must thank Mr D. Chitty for this reference.

* [As they are beginning to be: see the literature cited in note 1, p. 22.]

we call the senile state is in the ecologist's sense merely pathological. Senility is an artifact of domestication, something discovered and revealed only by the experiment of shielding an animal from its natural predators and the everyday hazards of its existence. In this sense, no form of death is less 'natural' than that which is commonly so called.

Some interesting conclusions may be drawn from the fact that the latter end of life is ecologically atrophic or vestigial. It has several times been pointed out[1] that the changes which an animal may undergo after it has ceased to reproduce are never directly relevant, and are in most cases quite irrelevant, to the course of its evolution. A genetic catastrophe that befell a mouse on the day it weaned its last litter would from the evolutionary point of view be null and void. This state of affairs is tacitly acknowledged in the celebrated half-truth that 'parasites live only to reproduce': the whole truth is that what parasites do *after* they reproduce is not on the agenda of evolution. The same applies to what may befall a mouse when it reaches the age of three, though in fact it never (or hardly ever) lives that long. We shall return to this point later. For the present it may be said that the existence of a post-reproductive phase of life is not causally relevant to the problem of ageing, for it is just that very ingredient of the ageing process—the decline and eventual loss of fertility—which it is our chief business to explain.

* * * * * *

What is the upshot of all this speculation? I think many biologists would agree that Weismann was in principle correct, and that the process of senescence in the individual and the form of the age-frequency distribution of death that mirrors it statistically have been shaped by the forces of natural

[1] Cf. G. G. Simpson, *Tempo and Mode in Evolution*, p. 183, Columbia University Press, 1944.

selection.* But before looking into this belief more closely, it will be as well to start this section, like its less technical predecessor, with a few definitions.

First, 'evolution'. Biologists often speak of organs, tissues and even cells 'evolving', but it must be recognized that this manner of formulation is by modern lights imprecise, or, what is not quite the same thing, inexpedient. These various things do indeed participate in evolution, just as our noses participate in our motion without themselves being mobile. What moves in evolution, what evolves, is an animal *population*, not an individual animal; and the changes that occur in the course of evolution, if we put a magnifying glass to them instead of feeling obliged to peer dimly down the ages of geological time, are changes in the composition of a population and not, primarily, in the properties of an individual. In visual analogy they are to be likened, not to a transformation scene at the pantomime, but to the sort of overlapping transformation we watch at the cinema when one 'set' slowly evaporates and is dispossessed of the screen by another.

Further, whatever form evolution may take, or whatever may bring it about, contributions to evolutionary change are paid, if they are paid at all, in one currency alone: offspring. Animals favoured by the process which, wise after the event, we call 'natural selection', make an extra contribution, however small, to the ancestry of future generations; and this brings about just that shift in the genetical composition of a population which we call an 'evolutionary change'. The problem of measuring natural selection, which so worried Karl Pearson,[1] is thus solved: the magnitude of natural selection is measured by the relative increase or decrease in the frequency with

[1] Cf. K. Pearson, *The Chances of Death and Other Studies in Evolution*, 2 vols., London, 1897.

* [The argument sketched in this section is developed more fully in the essay which follows, *An Unsolved Problem of Biology*.]

which the factor which governs some heritable endowment appears in the population.

I said, earlier on, that any theory of the origin of the ageing process must take two things into account: the early onset of what is in the technical sense senescence, and the continuity of its expression through life. I would like now to suggest that the 'force of mortality' has been moulded by a physical operator that has the dimensions of time × luck. Let us examine how natural selection will work upon a population that is potentially immortal; of which the individuals remain, for all the time that they are alive, in the fullness of physical maturity. Such a population will contain old animals and young. The old are old in years alone: we are so used to hearing the overtones of senility in the word 'old' that we must forcibly adjust ourselves to accept this important qualification. The old animals I shall speak about are 'in themselves' (to use a category of lay diagnosis) 'young'. They will no doubt have the advantage of their juniors in reflex and immunological wisdom, but these advantages will in the first approximation be disregarded.

Upon this population exempt from age decline we shall now superimpose a variety of causes of death that are wholly random or haphazard in their manner of incidence. The causes of death being random in nature, and susceptibility to it independent of age, it follows that the probability that an animal alive at the beginning of any span of time will die within its compass is likewise constant. The one-year-old is just as likely to see his second birthday as is the fifty-year-old to see his fifty-first. But the chances *at birth* of living to age 1 and age 50 are very different indeed; for as Weismann pointed out, though the significance of it escaped him, the older an animal becomes the more frequently is it exposed to the hazard of random extinction. Likewise a coin that has turned up heads ten times running will turn up heads on the eleventh spin in just 50 per cent. of trials; but the chances of turning up heads eleven times running are very small indeed. The upshot of this is that young animals will always outnumber old.

Let us in imagination mark a group of 100,000 animals at birth and follow it through life, supposing that the chance of dying within any small interval of time is constant, and equal to one-tenth per annum of those that remain alive to submit to the hazard. The survivors at the end of the first year will be 90,000; at the end of the second year, nine-tenths of those alive at its beginning, namely 81,000; and so on, through 72,900, to numbers which obviously get very small. In a population with a 'life table' such as this, supposing that it is not decreasing in numbers, a certain steady state of ages will be reached, a certain definite age-spectrum or composition with regard to age. At this steady stage, youngsters are being fed into the lower reaches of each age-group at the same rate as death and the passage of time remove them from it. The shape of this 'stable age distribution' (which is moulded, odd though it may seem, by the birth-rate per head alone) is that of a die-away exponential curve, such as one so often meets in the numerical treatment of natural data. The number of animals in each age-group bears a constant ratio, greater than unity, to the number of animals in the age-group following next.

What is important from our point of view is that the contribution which each age-class makes to the ancestry of future generations decreases with age. Not because its members become progressively less fertile; on the contrary, it is one of our axioms that fertility remains unchanged, so that the reproductive value *per head* is constant;[1] but simply because, as age increases, so the number of heads to be counted in each age-group progressively falls. It is at least as good a guess as Weismann made, that the process of senescence has been genetically moulded to a pattern set by the properties of this

[1] The term is technically defined in R. A. Fisher, *The Genetical Theory of Natural Selection*, Chap. 2, Oxford, 1930.

'immortal' age distribution. It is by no means difficult to imagine a genetic endowment which can favour young animals only at the expense of their elders; or rather, at their own expense when they themselves grow old. A gene or combination of genes that promotes this state of affairs will under certain numerically definable conditions spread through a population simply because the younger animals it favours have, as a group, a relatively large contribution to make to the ancestry of the future population. It is far otherwise with a genetic endowment which favours older animals at the expense of young. Reflection will show that the gene or genes concerned cannot plead for a retrospective judgement in their favour; for before the animals which bear these genes give outward 'phenotypic' evidence of the fact, they are on equal terms with those that do not. The greater part of the ancestry of the future population will thus have been credited indifferently to both types, because a gene qualifies for the preferential action of natural selection only when, to put it crudely, it manifestly works. This does not imply that a late-acting gene which confers selective advantage cannot spread through the population. It can indeed do so; but very much more slowly than a gene which gives evidence of itself earlier on. The later the time in life at which it appears, the slower will be its rate of spread; and the rate in the end becomes vanishingly small.

The consequence of any decline in the fertility of older animals is cumulative. Once it has happened, a new set of events may be put in train. Consider the fate of genetic factors that make themselves manifest in animals that bear them, not at birth nor in the first few days of life but at some time later on. Quite a number of such genes are known, and what is said of them applies equally to genes which have an expression, but a variable form of expression, throughout the whole span of life. It may be shown that if the time of action or rate of expression of such genes is itself genetically modifiable, then, if the gene confers selective advantage, its time of action or of optimal expression will be brought forward towards youth, as it spreads through the population. If, by contrast, the gene is 'disadvantageous', then its time of action or threshold of unfavourable expression will be pushed onwards in life while it is being eliminated from the population. The former process may be called a precession of favourable gene effects; the latter, a recession of unfavourable effects. Neither process can come into operation unless the fertility of the population declines with age, so that the reproductive value of its members falls; and the latter process, the recession of unfavourable gene effects, will be modified by the fact that the later an 'unfavourable' gene comes into operation, the slower will be the process of its removal from the population. (At some critical late age, perhaps, an unfavourable gene is eliminated so slowly that natural selection cannot challenge its reintroduction into the population in the process of gene mutation.) The precession of 'favourable' gene effects will in its turn be modified by the fact that reproduction cannot start at birth, and nature has found in higher animals only the most indirect substitutes (maternal care, and the blunderbuss of huge fecundity) for the theoretically desirable state of affairs in which an animal is born mature. Because of the hazards to which baby animals are exposed (and this is just as true of human beings) the reproductive value of the individuals always rises to a maximum before eventually it falls; and it is at the epoch of this maximum, therefore, that the 'precession' of favourable gene effects will automatically come to halt. It is not surprising, then, to find that in human beings the 'force of mortality' is lowest just when the reproductive value would in the members of a primitive society be highest—in the neighbourhood of the fourteenth or fifteenth years of life.[1] Nor is it surprising to find that

[1] A correlation pointed out by R. A. Fisher (see note 1, p. 37).

'senescence' begins then, rather than at the conventionally accepted age of physical maturity somewhat later on.

* * * * * *

The foregoing paragraphs represent no more than a few extra guesses woven in among Weismann's original hypothesis of ageing. If what Weismann believed is true, then nothing very radical can be done by way of modifying the course of growing old. Scientific eugenics could in the long run give us a more generous span of life; but only, it seems, by engaging life in lower gear, by piecing out the burden of the years into a larger number of smaller parcels, so prolonging youth symmetrically with old age. But the inevitability of old age does not carry with it the implication that old age must be a period of feebleness and physical decay. If specific secretions of the ductless glands fail; if assimilation becomes less efficient, so that essential food factors fail to penetrate the gut wall; if chronic low-grade infections persist because the defences of the body lack power to overcome them; in all such cases it should be possible to remove, at least for a while, any ingredients of the senile state for which they may be specifically responsible. The solution of these problems is a matter of systematic empirical research.

Side by side with research of this type there should be undertaken a thoroughgoing physiological analysis of the mechanism of ageing. I shall sketch one possible line of analysis here, because although the layman often understands the nature of scientific problems and can usually grasp the principles of their solution, he has, as a rule, very little idea of how scientific work is actually done.

If a physiologist were to study the problem of ageing from scratch, he would not even begin to try to modify the time-course of senescence by the administration of vitamins or elixirs compounded of the juices of the glands. He would first of all try to piece together a full empirical description of the phenomenon of ageing, as it is reflected in structural changes of tissues and cells and, more particularly, in the type and intensity of tissue and cellular metabolism. Only scraps of such information are now available: he would have to collect more. (The physiologist might in any case become more fully aware of the dimension of time in his experimental work. Nearly all his work is done with mature animals; studies on youngsters and animals past the reproductive period are far too few.)

With an adequate background of purely descriptive evidence, the physiologist could then bring the experimental method to bear. The first problem he would seek to solve is this: is the phenomenon of ageing something 'systemic' in nature—something manifested only by systems of the degree of organization of whole animals—or is it intrinsically cellular? Studies on tissue cultivation have given a partial answer to this question, but there are grounds for supposing that in certain critical respects it is misleading. One promising alternative that has become available to him is the technique of tissue and organ transplantation between animals of different ages. The majority at least of the members of very highly inbred strains of mice are from the standpoint of tissue-interchange genetically identical, for after many generations of repeated brother-to-sister mating they come to resemble each other (sexual differentiation apart) almost as closely as identical twins. One may therefore interchange parts of their bodies on a scale limited only by the exigencies of technique; one may make time-chimeras of youth and old age.* How, then, does tissue transplanted from a baby animal to a dotard develop in its 'old' environment? Does it rapidly mature and age, or does it remain like a new patch on an old pair of socks? Conversely,

* [The grafting of tissues between animals of different ages might be described as 'heterochronic', and it does for age or time what 'heterotopic' grafting does for place or space; see *The Uniqueness of the Individual*. Professor P. L. Krohn is making a particular study of these problems.]

what is the fate of tissue grafted from old animals into youngsters? If ordinary laboratory mice are used for such experiments, as very likely they have been, or even what are sometimes with undue optimism called 'pure strains', then the evidence is falsified at the outset; for the transplantation of tissues between animals very little dissimilar genetically simply provokes an immunity reaction, not different in principle from that which governs the outcome of certain blood transfusions, as a consequence of which the grafted tissue is destroyed.* But if suitable genetic precautions are taken, these problems and others of equally wide compass are capable of solution. Only when they are solved can the physiologist begin to ask more specific questions, such as whether the determinative factors of ageing are humoral in nature or of some other more complex type.

It is rather urgent that research of this type should be undertaken. Man's mean expectation of life at birth has increased very dramatically over the last 100 years, but chiefly as a consequence of reduced mortality in infancy and childhood. The mean expectation of life at the age of forty has increased hardly at all. But because of this reservation for life of many who would otherwise have died, the age-spectrum of the population, i.e. the proportion of its members within each age-group of life, is in many civilized countries shifting slowly towards old age. In forty years' time we are to be the victims of at least a numerical tyranny of greybeards—a matter which does not worry me personally, since I rather hope to be among their numbers. The moral is that the problem of doing something about old age becomes slowly but progressively more urgent. Something must be done, if it is not to be said that killing people painlessly at the age of seventy is, after all, a *real* kindness. Those who argue that our concern is with the preservation of life in infancy and youth, so that pediatrics must forever take

* [See *The Uniqueness of the Individual* herein.]

precedence of what people are beginning to call 'gerontology', fail to realize that the outcome of pediatrics is to preserve the young for an old age that is grudged them. There is no *sense* in that sort of discrimination.

2

An Unsolved Problem of Biology

The problem I propose to discuss is that of the origin and evolution of what is commonly spoken of as 'ageing'. It is a problem of conspicuous sociological importance. Everyone now knows that the proportion of older people in our population is progressively increasing, that the centre of gravity of the population is shifting steadily towards old age. Using a plausible combination of hypotheses, one among several, the Statistics Committee of the Royal Commission on Population predicts that in half-a-century's time one-quarter of our population will be not less than sixty years of age. The economic consequences of such an age-structure are all too obvious. Now biological research is by no means uninfluenced by the economic importunities of the times, and there can be little doubt that the newly awakened interest of biologists in ageing—or the hard cash that makes it possible for them to gratify it—is a direct reaction to this economic goad. Unfortunately, scientists have been slow to realize that the biologically important consequences of this secular increase in average longevity began to be apparent three-quarters of a century ago and are now on the threshold of completion. About seventy-five years ago, the mean expectation of life at birth in England and Wales began to exceed, as it now greatly exceeds, the age beyond which child-bearing virtually ceases. Women have had nearly all their children by the time they are forty-five, but may now expect, on the average, to live some quarter of a century longer. The fertility of men lasts beyond that of women and ends less sharply, but, roughly speaking, three-quarters of the male population is still alive at an age at which it can be credited with 99 per cent. of its children. The principal causes of death have changed accordingly. Fifty years ago the major killing diseases were pneumonia and tuberculosis, both of infective origin; to-day they are cancer and what is compendiously called cardiovascular disease. Susceptibility to both cancer and the cardiovascular diseases is in some degree influenced by heredity, and should therefore be subject to those forces, of 'natural selection', that discriminate between the better and the genetically less well endowed. (To speak of 'discrimination' is, of course, to put the matter in too literary a way; let us say that people with different hereditary endowments do not have children in strict proportion to their numbers; some of them take more than their numerically fair share of the ancestry of future generations.) But cancer and the cardiovascular diseases are affections of middle and later life. Most people will already have had their children before the onset of these diseases can influence their candidature for selection. In the post-reproductive period of life, the direct influence of natural selection has been reduced to zero[1] and the principal causes of death to-day lie just beyond its grasp.

[1] The word 'direct' is important. Grandparents, though no longer fertile, may yet promote (or impede) the welfare of their grandchildren, and so

How it is that the force of natural selection becomes attenuated with increasing age I hope to explain very fully later. What is important in the meantime is that one should realize how, in the last seventy-five years, the whole pattern of the incidence of selective forces on civilized human beings has altered. We are not now waiting for our ageing population to produce biological changes of first-class importance, as some demographers seem to suggest. The changes have already happened. We have already entered a new era in the biological history of the human race.

II

It is a curious thing that there is no word in the English language that stands for the mere increase of years; that is, for ageing silenced of its overtones of increasing deterioration and decay. At present we are obliged to say that Dorian Gray did not exactly 'age', though to admit that he certainly grew older. We obviously need a word for mere ageing, and I propose to use 'ageing' itself for just that purpose. 'Ageing' hereafter stands for mere ageing, and has no other innuendo. I shall use the word 'senescence' to mean ageing accompanied by that decline of bodily faculties and sensibilities and energies which ageing colloquially entails. Dorian Gray aged, but only his portrait

influence the mode of propagation of their genes. A gene for grandmotherly indulgence should therefore prevail over one for callous indifference, in spite of the fact that the gene is propagated *per procurationem* and not by the organism in which its developmental effect appears. Selection for grandmotherly indulgence I should describe as 'indirect', and the indirect action of selection becomes important whenever there is any high degree of social organization. The genes that make for efficient and industrious worker bees, for example, are of vital importance to the bee community, though not propagated by the worker bees themselves. Dr Kermack points out that the distinction between 'direct' and 'indirect' selection can easily be misleading, because in the outcome their effects are both the same. Let us admit, however, that there is a distinction of genetical procedure, though it might well have been embodied in better-chosen terms.

disclosed the changes of senescence. I hope that makes it clear.

Senescence means a decline of vitality. How is this to be more precisely defined and measured? One may set about trying to measure senescence in two entirely different sorts of ways.

The first sort of measure is personal, in the sense that it is carried out on individual animals. Quite a number of schemes of measurement are at our disposal. For example, the rate at which wounds heal provides some sort of measure of what we vaguely mean by vitality, since it depends on the multiplication or migratory activity of cells. What sort of answer does it give? So far as we know, the answer is that the rate of wound healing is highest at birth and steadily declines thereafter. In terms of this measurement, therefore, senescence begins at birth and the 'prime of life' is something of a fiction. Or we might reasonably choose a measure founded on the acuity of the senses. The acoustical prime of life, for example, appears to be in the neighbourhood of the age of ten, for we are said to hear sounds of higher pitch at ten than earlier or thereafter. On the other hand physical strength, endurance and the niceties of muscular co-ordination reach their peak at about twenty-five.

All these are very piecemeal measures. The best, perhaps, is that originally devised by Minot—the multiplicative power of the tissues of the body, that is, their capacity to increase by further growth in the manner in which they themselves were formed. Organisms tend to grow by compound interest, for that which is formed by growth is itself usually capable of further growing. But the rate of interest falls; the organism grows like a sum of money which, invested at birth at (say) 10 per cent. compound interest, gathers in a lower rate of interest year by year. The rate of interest does indeed fall from birth, and it is at birth, if Minot is to be believed, that senescence must be said to begin. And so, in some perfectly respectable sense, it does; but if we pursue this train of thought by asking in what manner the rate of interest falls, we shall be led by Minot into an attractive paradox. The answer is that from birth onwards the rate of interest falls steadily at a rate which itself steadily falls. Not only does senescence begin at birth, but it is going on much faster in the early years of life than latterly. The child is hurrying precipitately towards his grave; his elders, appropriately enough, proceed there in a more decent and orderly fashion.

None of these personal measures is of more than limited value. They are together incomplete, and severally give different answers; nor can they be made to add up to give a single figure that represents a measure of senescence in the round. Let us therefore turn to a scheme of measurement founded on wholly different principles.

III

The second sort of measure is not personal, but statistical. We have agreed that senescence is a decline of what may be vaguely called vitality, and must now ask what property it is that changes as a direct outcome of that decline. The property is, in a word, *vulnerability* to all the mortal hazards of life; and it is measured by the likelihood of dying within any chosen interval of age.

The measurement of vulnerability is in principle very easy. Imagine 100,000 animals, each of which is labelled or otherwise identified at birth and followed throughout its life; and suppose one keeps a record of the age at which each dies, keeping the record open until the death of the most long-lived. Such a record might well be called a Death Table, but, by an agreeable euphemism, it is in fact called a Life Table. If we plot the number of survivors against age, the curve so defined starts with age 0 at 100,000 and falls to zero at the age of about 100 years. Fig. 1 illustrates the shape of the life-table curve for human beings.

From such a curve one may compute the death-rate at any age of life, for that is simply its slope, the rate of decline of the number of survivors; the mean expectation of further life at birth or at any other age; and the likelihood at any one age of living

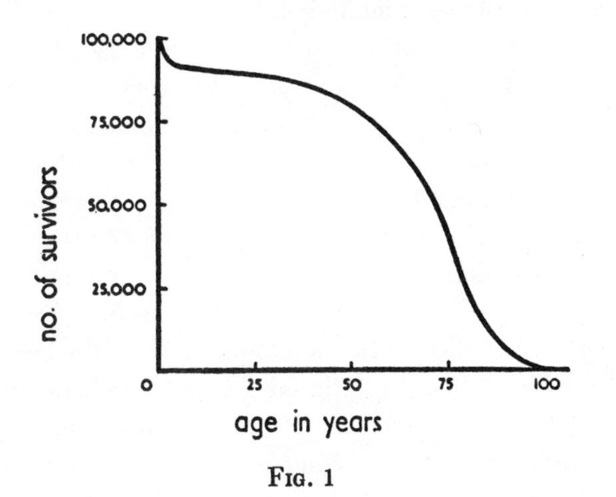

FIG. 1

to any other. The property that concerns us, however, is that which is called the specific death rate or, less aridly, the 'force of mortality', the likelihood of dying within each interval of age. In a first approximation, which is all that is necessary for our purpose, the force of mortality is the quotient of this fraction:

$$\frac{\text{Number of organisms that die within any chosen interval of age}}{\text{Number of organisms alive at the beginning of the interval}}$$

If, for example, 100 men reach age eighty-nine, but only 80 of them reach age ninety, then the force of mortality in the ninetieth year of life is simply 0.2 (20 per cent., or 200 in every 1000). If there is no senescence in the population—if vitality does not decline, so that there is no greater likelihood of dying at any one age than at any other—then the force of mortality must necessarily be constant. Its members die, to be sure; but a man who has just celebrated his eightieth birthday anniversary is no more or less likely to celebrate his eighty-first than is a seven-year-old to celebrate his eighth. In my diagram, the force of mortality, being constant, would appear as a straight line parallel to the axis defining age.

In real life it is far otherwise. As Fig. 2 shows, the vulnerability of newborns is, not unexpectedly, very high; not until nearly the seventieth year of life does it become so high again. The curve of the force of mortality falls precipitously to a minimum around age twelve and then climbs upwards, slowly at first and latterly much faster. Age twelve (or thereabouts) is therefore the actuarial prime of life; at twelve one is more likely than at any other age to survive one further year, or month, or minute. But notice the smoothness of the curve that defines the force of mortality in later life. There is no break or singularity to give evidence that at any later age development and maturation are at least completed and that deterioration then sets in. Any complete theory of the origin and evolution of senescence must explain the smoothness and coherence of the curve of increasing vulnerability. It is not quite good enough merely to think up reasons why very elderly animals should die.

Because there are clearly special reasons why baby animals should be more vulnerable, though no less charged with vitality than their elders, I am proposing to neglect the arc of the curve of the force of mortality that lies before its minimum, but to use its later stretch as a measure of the degree of senility. This is a decision that cries aloud for qualifications and reservations, and it is part of my purpose to reveal what some of these may be.

You will notice first that although the force of mortality may purport to measure a process that happens in the life of an individual animal (decline of 'vitality', or what you will) it is in fact founded upon the age-frequency distribution of a single event in life—its end. It is a notorious fact that Maxwell's Demon can reduce all such measures to absurdity, since he can strike down perfectly vigorous, or indeed potentially immortal, animals at just such ages as will exactly imitate any chosen force-of-mortality curve.

There are many other serious reservations. The use of the force of mortality as a measure of senescence assumes that all

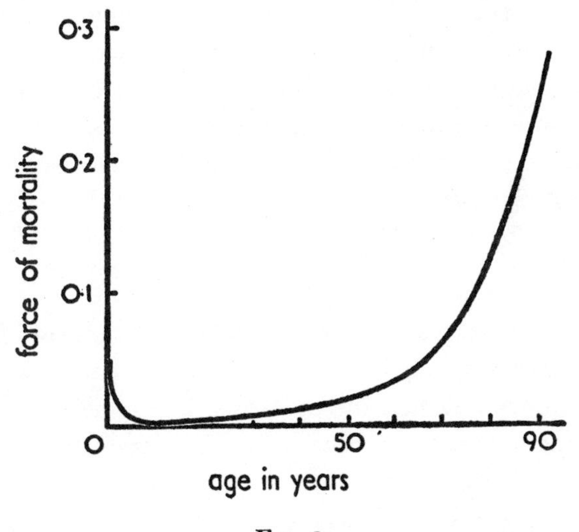

Fig. 2

members of the population are equally at risk. This is not true, because wage-earners are more exposed to risk than school-children or those who have retired. A third difficulty is that if a life table is constructed in the way I have suggested—that is by following the life histories of a cohort of the newly born—it is only too likely to be corrupted by secular changes in the hazards of which human beings may be victims. Individuals aged seventy to-day were born in 1881, when the causes of the death of children, and their likelihood of surviving early youth, were very different from what they are to-day. A fourth difficulty is that if the population is rather crudely subdivided into the innately (that is, genetically) less tough and tougher, then the population that reaches age sixty will be by no means a genetically fair sample of the cohort with which the life table began. Presumably each pattern of genetic constitution endows its owners with a characteristic mode of increase of vulnerability; but in a cohort of mixed origins all such distinctions must inevitably be confused.

These are grave difficulties,* but all of them can be overcome in principle, and some in laboratory practice. I now turn to a much more important difficulty in the use of vulnerability as a measure of senescence; it is ingrained, and in practice ineradicable, and it leads us to distinguish between two *sorts* of causes of senescence.

IV

Consider wrinkles and lines on the skin, for these are familiar outward signs of ageing in its colloquial sense. People who often frown get lines between the eyebrows; the supercilious reveal their temperament by furrows across the forehead; deep lines down the corners of the mouth are allegedly the consequence of having a ready smile. What is the history of wrinkles? Every time one grins or frowns some physical trace is left in the cellular or fibrous structure of the skin. These traces are cumulative, and if only one folds or creases the skin sufficiently often, they will add up to form a visible flexure line. One perfectly good reason why elderly people should have more lines and

* [There is another difficulty in accepting 'vulnerability' as a measure of senescence: the decline and loss of reproductive power (e.g. in the menopause) is beyond question a form of senescence, but it is not accompanied by any increase of vulnerability in an actuarial sense. I consider this problem, and deal more fully with the other difficulties mentioned in the text, in *The definition and measurement of senescence*, Ciba Foundation Colloquia on Ageing, Vol. 1, p. 4, 1955.]

wrinkles is therefore simply that, being older, they have frowned and grinned more often. But we must also ask whether the skin of older people more readily takes the impress of creasing and folding. Does a *first* flexure in the skin of an older person leave a bolder trace than a first flexure in the skin of someone younger? We may be certain that it does. But the point is that *both* an increase in innate susceptibility to wrinkling *and* the cumulative effect of recurrent creasing have played a part in the history of wrinkles; and although we can distinguish the two sorts of causes in theory and in experimental practice, they cannot be disentangled merely by contemplating the wrinkle as a *fait accompli*.

Wrinkling is an unimportant example of a kind of disability that affects all animals. Any injury that leaves a physical trace, as all but the most trivial do, increases the vulnerability of older animals, because injuries of one sort or another are recurrent hazards and older animals, having been exposed to them more often, will have built up a bigger actuarial debt. Skin scars may be individually trivial things, but the older animals will have more of them; and apart from that, germs gain easier access to the body during the time taken by a wound to heal. Fractures of bone are slow to reunite and animals make easy prey until they have done so. The heightened blood-pressure that accompanies the shocks and alarms of natural living predisposes the blood-vessels to degenerative change. Cells may produce faulty copies of themselves in what should be an act of exactly symmetrical division; division is recurrent and faulty copies are perpetuated, so that their ill effects, summed over the cell population of the body, are bound to add up. The efficacy of most of the known cancer-provoking chemical compounds depends upon the repeated exposure of tissues to their action over long periods. Infections are recurrent hazards; most infections damage cells, and some do permanent damage of a sort that increases vulnerability in an obvious way.* To go back to colloquial speech, all these effects are the effects of age but not necessarily the effects of ageing; they may take their toll even if ageing is not accompanied by an innate deterioration. Senescence, as it is measured by increase of vulnerability or the likelihood of an individual's dying, is therefore of at least twofold origin.[1] There is (*a*) the innate or ingrained senescence, which is, in a general sense, developmental or the effect of 'nature'; and (*b*) the senescence comprised of the accumulated sum of the effects of recurrent stress or injury or infection. The latter is environmental in origin and thus, paradoxically, the effect of 'nurture'. There is always an empirical test for distinguishing between the two in principle—one has only to find out whether a *first* injury or physiological abuse or stress is less well tolerated by older animals than their juniors—but in the actual records of vulnerability the two are inextricably combined.[2]

* [Mammals which have what is optimistically described as a 'permanent dentition'—i.e. a second and final set of teeth—obviously depend upon its remaining in working order; but teeth are bound to get chipped or damaged in the ordinary course of biting, and this is a good example of deterioration of the kind classified below under heading (*b*).]

[1] Dr Whitear has pointed out that a third and quite distinct sort of change with ageing which influences and will ultimately increase the vulnerability of older animals is that entailed by the differential growth and changing proportions of the several organs of the body or ingredients of a complex tissue. As a general rule, it may be said that every fixed regimen of differential growth will, if growth is indeterminate, inevitably lead to mechanical or physiological ineptitude of one sort or another, although not necessarily involving a loss of 'vitality' at the cellular or tissue level. The problem is discussed more fully later.

[2] Higher organisms have means for counteracting the cumulative effect of recurrent injuries. Two of the three principal reflex (i.e. response-to-stimulus) systems of the body, the immunological and the nervous, have the power of 'storing their information' for long periods. The hormone system, apparently, has not. In general, an animal is less likely to contract a particular infection on its second exposure than on its first, and this is mainly due to the fact that what immunologists call the 'secondary' response to an immunity-provoking agent is a good deal brisker than the first. An animal is also less likely to get bitten, burnt, or otherwise abused

That one is obliged by the terms of my definition to admit that there are two sorts of causes of senescence has, it will turn out, no more than a minor nuisance value. I am of course chiefly concerned with senescence of sort (*a*), and you will see that the arguments put forward to account for its origin and evolution are greatly strengthened by the fact that there may already exist a senescence of sort (*b*).

The time has now come for a formal definition of senescence, and I shall adopt the usual practice of translating a statement about the frequency of the occurrence of an event in a population into a statement about the likelihood of its happening to an individual. Senescence, then, may be defined as that change of the bodily faculties and sensibilities and energies which accompanies ageing, and which renders the individual progressively more likely to die from accidental causes of random incidence. Strictly speaking, the word 'accidental' is redundant, for all deaths are in some degree accidental. No death is wholly 'natural'; no one dies *merely* of the burden of the years.

V

By way of an interlude let me now, as a zoologist, apologize for appealing so much to evidence from human beings. I do so because we know so very much more about the death of human beings than of other animals; and though I feel a professional obligation to say something about the natural history of senescence, there is no time to do so, and even if there were, there would not be much to say.

at each successive exposure to such a hazard; it will have 'remembered' the earlier and accordingly learnt better. Two exposures to infection or physical risk may therefore have a no more harmful consequence than one, and the cumulative effects of some sorts of recurrent stress may therefore be to some extent corrected by the benefactions of an immunological or nervous memory.

We can be quite sure that mammals undergo a process of 'innate' senescence. But why are we so sure? The answer is vital to my later argument. It is because we keep mammals as pets, in zoos, and in domestication. If we had to rely upon information derived from truly wild animals, we should be very much indeed less certain, and it is arguable that we might never know at all. For wild mammals of any perceptible degree of senility turn up in traps so seldom that we should always be inclined to think up reasons for their enfeeblement that were not necessarily connected with their age—the wasting due to infection, maybe, or to an injury that stopped them getting food. Animals do not in fact live long enough in the wild to disclose the senile changes that can be made apparent by their domestication. Many wild birds, as Dr Lack has shown, are the victims of so savage an exaction of mortality that, beyond a few months of youth, their likelihood of dying is actually independent of their age! It is of vital importance to remember that senility is in a real and important sense an artifact of domestication; that is, something revealed and made manifest only by the most unnatural experiment of prolonging an animal's life by sheltering it from the hazards of its ordinary existence. Here is a story with a pertinent moral. An eminent naturalist was once taken tiger-hunting by a courteous Indian potentate; he got his tiger and saw at once that it was very, very old. Here then perhaps, he thought, is something that he had long vainly looked for—a truly wild animal that was very old and very decrepit, and no doubt very cunning and very wise as well. On closer inspection, he found that the tiger had gold fillings in its molars; the potentate, courteous as I said, had simply 'laid it on'. So when you hear speak of the 'natural death' of animals, remember that no death is less 'natural' than that which is commonly so called.

If there are doubts about mammals and birds, which represent the higher classes of vertebrates, how many more must there be about the members of what we are now obliged to call the under-privileged classes? There is still, it appears, no more to be said about senescence in fish than was said by my predecessor Sir Edwin Ray Lankester some eighty years ago: 'Fish are not known to get feeble as they grow old, and many are known *not* to get feebler.' My professional colleagues will know that Dr G. P. Bidder held some fascinating and far from implausible views on the origin of senescence which turn on the belief that fish do not deteriorate with ageing. These I cannot delay with. But is it not a most revealing fact that there should be any doubt about the matter at all? Fish *may* be potentially immortal in the sense that they do not undergo an innate deterioration with ageing, and yet the naturalists who ought to know about it simply can't be sure! As you will see, this uncertainty is the most tell-tale evidence in favour of my later argument. Whether animals *can*, or cannot, reveal an innate deterioration with age is almost literally a domestic problem; the *fact* is that under the exactions of natural life they do not do so. They simply do not live that long.

VI

I have deliberately spent more than half my time in discussing the measurement and definition of senescence, and I now want to discuss the factors that may have played their part in its origin and evolution. As a text I shall use a quotation from the works of August Weismann.

> Death takes place because a worn-out tissue cannot for ever renew itself. Worn-out individuals are not only valueless to the species, but they are even harmful, for they take the place of those which are sound . . . by the operation of natural selection, the life of a theoretically immortal individual would be shortened by the amount which was useless to the species.

Weismann's propositions have the great merit of suggesting,

for only the second time, that senescence has had a very orthodox evolutionary origin. But Weismann is arguing in what a student of mine once called a viscous circle, or more exactly a vicious figure-of-eight. He assumes that the elders of his race are worn out and decrepit—the very state of affairs whose origin he purports to be inferring—and then proceeds to argue that because these dotard animals are taking the place of the sound ones, so therefore the sound ones must by natural selection dispossess the old! This is all a great muddle, but there is certainly some truth in it, and I shall spend the rest of my lecture in an attempt to find out what that truth may be.

My argument starts with a discussion of certain demographic properties of a population of potentially immortal individuals, and it will be illustrated by an inorganic model which I shall animate step by step. This choice makes it possible to avoid two common traps. The first of these is to argue that senescence in higher animals has come about *because* they have a post-reproductive period; for 'unfavourable' hereditary factors that reveal their action only in the post-reproductive period are exempt from the *direct* effects of natural selection and there is therefore little to stop them establishing themselves and gaining ground. Any such argument is wholly inadmissible. The existence of a post-reproductive period is one of the consequences of senescence; it is not its cause. The second trap, into which Weismann fell headlong, is to suppose that a population of potentially immortal individuals subject to real hazards of mortality consists in high proportion of very aged animals with a relatively small number of no doubt browbeaten youngsters running round between their feet. It will soon be clear that this idea is equally mistaken.

I want you now to consider a population of objects, living or not, which is at risk—in the sense that its members may be killed or broken—but which is potentially immortal in the sense that its members do not in any way deteriorate with ageing. Test-tubes will do, since they are clearly 'mortal', and I shall peremptorily assume that they do not become more fragile with increasing age.*

Imagine now a chemical laboratory equipped on its foundation with a stock of 1000 test-tubes, and that these are accidentally and in random manner broken at the rate of 10 per cent. per month. Under such an exaction of mortality, a monthly decimation, the activities of the laboratory would soon be brought to a standstill. We suppose therefore that the laboratory steward replaces the broken test-tubes monthly, and that the test-tubes newly added are mixed in at random with the pre-existing stock. The steward will obviously be obliged to buy an average of 100 test-tubes monthly, and I am going to assume that he scratches on each test-tube the date at which he bought it, so that its age-in-stock on any future occasion can be ascertained.

Now imagine that this regimen of mortality and fertility, breakage and replacement, has been in progress for a number of years. What will then be the age-distribution of the test-tube population; that is, what will be the proportions of the various groups into which it may be classified by age? The answer is illustrated in Fig. 3. The population will have reached the stable or 'life-table' age-distribution in which there are 100 test-tubes aged 0-1 month, 90 aged 1-2 months, 81 aged 2-3 months and so on. This pattern of age-distribution is characteristic of a 'potentially' immortal population, i.e. one in

* [In real life, of course, test-tubes could undergo senescence of both the types, (*a*) and (*b*), which I have distinguished in the text. 'Innate senescence' might be represented by the slow crystallization of the glass, which will happen whether the test tubes are used or not, and 'traumatic senescence' by the accumulation of tiny chips or cracks which, without making the test-tube unusable, make it a good deal more likely to be broken in everyday use. A life table for glass tumblers has been worked out by G. W. Brown and M. M. Flood, *Journal of the American Statistical Association*, **42**, p. 562, 1947.]

which the chances of dying do not change with age. The curve it outlines is of a sort very familiar in science. Fig. 3 illustrates this very elementary triusm: the older the test-tubes are, the

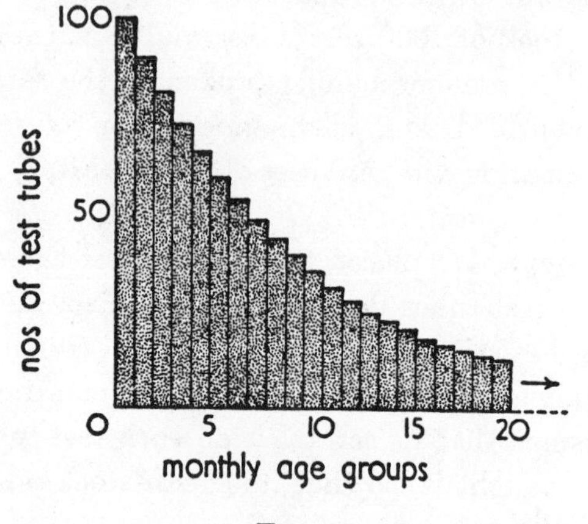

FIG. 3

fewer there will be of them—not because they become more vulnerable with increasing age, but simply because the older test-tubes have been exposed more often to the hazard of being broken. Do not therefore think of a potentially immortal population as being numerically overwhelmed by dotards. Young animals outnumber old, and old animals those still older.

VII

As a first step in animating this model, I want you to imagine that the test-tubes now do for themselves exactly what the steward has hitherto been doing for them, i.e. they reproduce themselves, no matter how, at an average rate of 10 per cent. per month in order to maintain their numbers. Since the population is potentially immortal, the rate of reproduction of its members will not vary with their age. It follows that each 'living' test-tube of the existing population will make the same average contribution of offspring to the test-tube population of the future. Each test-tube may lay claim to an equal share of the ancestry of future generations, and its reproductive value is invariant with its age.[1]

The next step in the argument is vital. Although each individual test-tube takes an equal share of the ancestry of the future population, each age-group most certainly does not. The older the age-group, the smaller is its overall reproductive value. The group of test-tubes 2-3 months old, for example, makes a very much greater contribution than the group 11-12 months old. This is not because the test-tubes of the senior group are individually less fertile—their fertility is *ex hypothesi* unchanged—but merely because there are fewer of them; and there are fewer of them not because they have become more fragile—their vulnerability is likewise unaltered—but simply because, being older, they have been exposed more often to the hazard of being broken. It is simply the old story of the pitcher and the well.

Some of the consequences of this decline in the reproductive

[1] The actuarial characteristics of a 'potentially immortal population' are particularly simple: the life table is defined by the relation $l_x = l_0 e^{-\mu x}$, where l_0 is the size of the original cohort, l_x is the number of them that survive to the age of x, and μ is the force of mortality $\left(\mu = -\frac{1}{l_x}\frac{dl_x}{dx}\right)$, independent *ex hypothesi* of age. The probability p_x of surviving from birth to age x is simply $l_x/l_0 = e^{-\mu x}$. If the number of offspring born to each member of the population in each unit of age remains constant, as we have supposed, at the value b, then the reproductive value remains constant throughout life at the value $R_x = \frac{1}{p_x}\int_x^\infty bp_x \, . \, dx = \frac{b}{\mu}$; and this will also be its value at birth (the net reproduction ratio R_0). If the regimen of constant mortality and fertility has been in progress long enough, and numbers are not declining ($b > \mu$), then a stable age-distribution will be reached in which the fraction of the population falling within the age interval x to $x + Dx$ is given by $c_x = \int_x^{x+Dx} be^{-bx}\, dx$; the proportion of the population aged x and upwards is thus simply e^{-bx}.

value of older age-groups will be apparent when I take the next step in animating my test-tube model. The test-tubes are no longer to be thought of as immortal; on the contrary, after a certain age, as a result of some intrinsic shortcoming, they suddenly fall to pieces. For the time being we shall assume that they disintegrate without premonitory deterioration. What will be the effect of this genetically provoked disaster upon the well-being of the race of test-tubes? It must be my fault if the answer does not appear to be a truism—that it depends upon the age at which it happens. If disintegration should occur five years after birth, its consequences would be virtually negligible, for under the regimen which we have envisaged less than one in five hundred of the population is lucky enough to live so long. Indeed, if we relied upon evidence derived solely from the natural population of test-tubes, we should probably never be quite certain that it really happened. We could make quite certain, as we do with animals, only by domesticating our test-tubes, shielding them from the hazards of everyday usage by keeping them in a padded box as pets.

If disintegration should occur one year after birth, an age which is reached or exceeded by about one-quarter of the population, the situation would be fairly grave but certainly not disastrous; after all, by the time test-tubes have reached the age of twelve months they have already made the greater part of their contribution of offspring to the future population. But with disintegration at only one month, the consequences would obviously be quite catastrophic.

This model shows, I hope, how it must be that the force of natural selection weakens with increasing age—even in a theoretically immortal population, provided only that it is exposed to real hazards of mortality. If a genetical disaster that amounts to breakage happens late enough in individual life, its consequences may be completely unimportant. Even in such a crude and unqualified form, this dispensation may have a real bearing on the origin of innate deterioration with increasing age. There is a constant feeble pressure to introduce new variants of hereditary factors into a natural population, for 'mutation', as it is called, is a recurrent process. Very often such factors lower the fertility or viability of the organisms in which they make their effects apparent; but it is arguable that if only they make them apparent late enough, the force of selection will be too attenuated to oppose their establishment and spread. Such an argument may have a particular bearing on, for example, the occurrence of spontaneous tumours and the senile degenerative diseases in mice of which Dr Gorer has made a special study, for these affections make themselves apparent at ages which wild mice seldom, perhaps virtually never reach. We only know of their existence through domestication; small wonder if they have no effect on the well-being of mouse populations in the wild. Mice, of course, do already show evidence of deterioration in the course of ageing, but my reasoning does not presuppose it. It applies to 'potentially immortal populations' with only a quantitative loss of cogency.

It is a corollary of the foregoing argument that the postponement of the time of overt action of a harmful hereditary factor is equivalent to its elimination.[1] Indeed, postponement

[1] As an example of what I mean by the time of 'overt action' of genes, I should say that the earliest age of overt action of a 'coat colour' gene was with the growth of a coat of hair in mice, which are born naked, or with birth in animals like the guinea-pig, which are born with a pelt of hair. It is not until hairs are both formed and exposed to outward inspection that the various coat colours, as such, can influence the welfare of their possessors. But I agree with Dr Grüneberg that one must be very cautious in speaking of the time of action of genes—for one important reason among several, because its influence on coat colour may be only one, and by no means the most important one, of the manifold actions of what is only for convenience of labelling described as a 'coat colour' gene. We have furthermore only the vaguest idea of what we mean by speaking of a gene's 'acting' at all. This particular difficulty can be overcome by accurate formulation:

may sometimes be the *only* way in which elimination can be achieved; but I cannot argue this without an appeal to the phenomena of pleiotropy and linkage, which time will not allow.

VIII

It is not good enough to say that what happens to very old animals hardly matters and that what happens to youngsters matters a great deal. For the degree to which anything may matter varies in a predictable way with age, and the selective advantage or disadvantage of a hereditary factor is rather exactly weighted by the age in life at which it first becomes eligible for selection. A relatively small advantage conferred early in the life of an individual may outweigh a catastrophic disadvantage withheld until later.[1] Go back to the test-tube model for a moment, and compare two competing test-tube populations. Both suffer the same average monthly mortality of 10 per cent., and one has, as hitherto, the average monthly birth-rate of 10 per cent. The other population has an average monthly birth-rate of 11 per cent., but the price paid for this hardly profligate increase of fecundity is the spontaneous bursting asunder of each member at age two. Which population will increase the more rapidly in numbers—the potentially immortal, or the mortal population with a birth-rate only

the time of action of a gene *G* with respect to a character *C* is the age at which, in a stated genetic and environmental context, the substitution of *G* for its allelomorph *G′* transforms the character *C′* into the character *C*. In short, it doesn't matter when (or even whether) *G* and *G′* are 'acting' until they give evidence of acting in different ways.

[1] By something that is a catastrophic disadvantage to an older animal I mean a change which is personally catastrophic, and which would certainly be catastrophic to the species as well if it made its appearance in younger animals. But in the strict sense, the verdicts 'advantageous' and 'disadvantageous' can be delivered only after trial by selection, and in this sense to speak of 'catastrophic disadvantages' which don't in fact much matter is self-contradictory.

one-tenth part higher than the other's? The simplest calculations show that it is the latter.

A heightened juvenile rate of reproduction, achieved perhaps at the expense of recurrent stress that later leads to deterioration, is by no means the only possible realization of the phenomenon illustrated by this model. It is a general rule, for example, that the parts of the body multiply their substance at unequal rates, so that proportions change as the body grows. There is very likely to be a 'best' proportion, or a best range of proportions, from the standpoint of functional efficiency and therefore of survival. In theory these proportions could be arrived at once and for all by starting the baby or embryo off with the appropriate shape and allowing growth to proceed by symmetrical enlargement. This does not happen in practice, and it is not biologically feasible for a whole variety of reasons. In practice, as I have already said, adult proportions are achieved by the adoption of a more or less fixed regimen of differential growth, i.e. of a more or less constant ratio between the multiplication rates of the several parts of the body. The danger inherent in this alternative solution is that there may well come a size, and therefore an age, at which proportions become functionally and structurally grotesque. The size of the male fiddler crab's claw increases as a power, greater than unity, of the size of the rest of its body, and Julian Huxley, who made a special study of these differential growth phenomena, points out that a crab whose body weighed 1 kg. would carry a claw about ten times that weight. But the sense of my argument is that if the appropriate proportions are achieved at some earlier stage of life, it may not much matter that the regimen of differential growth that brought them into being should eventually lead to mechanical ineptitude of this degree. The early advantage more than makes good the later disadvantage which it necessarily entails.

IX

The postponement of the time of overt action of 'unfavourable' hereditary factors is not just a good idea which the organism would be well advised to apply in practice; postponement may be enforced by the action of natural selection and senescence may accordingly become a self-enhancing process. Let me give you a real example in which this process appears to be happening at the present time.

Huntington's chorea is a grave and ultimately fatal nervous disability distinguished by apparently compulsive and disordered movements akin to, and perhaps identifiable with, 'St Vitus' Dance'. Its first full clinical description is in George Huntington's own memoir of 1872, though the evidence I shall appeal to comes largely from the fine treatise of Dr Julia Bell. Huntington's chorea is a hereditary affliction of a rather special sort. Its disabling and clinically important effects first become manifest not in youth or old age but at an intermediate period, its time of onset—later in men than in women—being most commonly in the age-group 35-39. Its age of onset does however vary, and I want you to assume (what is almost certainly true, though it would be hard to collect the evidence for it) that its age of onset, like the disease itself, is also genetically determined.

If differences in its age of onset are indeed genetically determined, then natural selection *must* so act as to postpone it: for those in whom the age of onset is relatively late will, on the average, have had a larger number of children than those afflicted by it relatively early, and so will have propagated more widely whatever hereditary factors are responsible for the delay. But as the age of onset approaches the end of the reproductive period, so the direct action of selection in postponing it will necessarily fade away.

One may now ask why, if such a thing must happen, has it

not happened already, and, if it has not, what is the evidence that it is happening now? The first question amounts to asking why Huntington's chorea is not *already* one of the diseases of the post-reproductive period, since selection of the sort I have outlined must be pretty vigorous and has presumably had tens of thousands of years at its disposal. My answer to this is based on an aside of J. B. S. Haldane's. It is only in the last century or so that selection has had a real chance to get a grip on it, for it is only within this period that the average expectation of life at birth has come to equal the average age of onset of the disease.* Even so, there is indirect evidence of a postponement of its age of onset. Since the male reproductive span is longer than the female's, the force of selection on men must be less quickly attenuated with increasing age; postponement should therefore have gone farther in men than in women—and this, as I have already said, is indeed the case. Ultimately, no doubt, the age of onset will come to a standstill in both men and women at the end of their respective reproductive periods. I gratefully acknowledge the origin of this train of thought in L. S. Penrose's writings on mental disease and natural selection.

With Huntington's chorea as a lucky concrete example, I can now propound the following general theorem. If hereditary factors achieve their overt expression at some intermediate age of life; if the age of overt expression is variable; and if these variations are themselves inheritable; then natural selection will so act as to enforce the postponement of the age of the expression of those factors that are unfavourable, and, correspondingly, to expedite the effects of those that are favourable —a recession and a precession, respectively, of the variable age-effects of genes. This is what I mean by saying that

* [This is not quite fair. It is not the mean expectation of life *at birth* that is important, but the mean expectation of further life at an age when reproduction has just begun. This too has increased, but not nearly so dramatically, over the past hundred years.]

senescence is a self-enhancing process. The theorem in the form in which I have just put it does not depend upon the existence of a post-reproductive period; it only requires that the reproductive value of each age-group should diminish with increasing age. I have argued that this must necessarily diminish even with a population of potentially immortal and indeterminately fertile individuals, provided only that they are subject to real dangers of mortality. In such a population a younger age-group must necessarily outnumber an older, for the older represents the residue of those who have been longer exposed to mortal hazards. If you should have, as I believe, unjustified qualms about an argument based upon combining an innate potential immortality with a contingent real mortality, I would recall to you my earlier distinction between senescence of sorts (*a*) and (*b*). Senescence of sort (*b*) is not innate or 'laid on' developmentally; it represents the outcome of the cumulative effects of recurrent physical damage, physiological stress, or faulty cellular replication. If you will admit that senescence of this sort is a means by which, irrespective of any genetical background, the reproductive value of each individual in a population is caused to diminish with increasing age, then my argument is quantitatively strengthened, because the numerical preponderance of the younger age-groups will become so much the more pronounced. And if, further, a post-reproductive period of life is already established, then indeed it becomes, as it were, a dustbin for the effects of deleterious genes. But these propositions are mere glosses or refinements. The argument must stand or fall on the case which I first proposed.

X

I have now suggested three agencies which may have played a part in the evolution of 'innate' senescence: (1) the inability of natural selection to counteract the feeble pressure of repetitive mutation when the mutant genes make their effects apparent at ages which the great majority of the members of a population do not actually reach; (2) the fact that the postponement of the time of action of a deleterious gene is equivalent to its elimination, and may sometimes be the only way in which elimination can be achieved; and (3) the fact that natural selection may actually enforce such a postponement, and, conversely, expedite the age of onset of the overt action of favourable genes. All these theorems derive from the hypothesis that the efficacy of natural selection deteriorates with increasing age.

I am inclined to think that the third factor, the enforced precession and recession of the ages of the overt action of genes, has the widest ambit of significance. But although I have foresworn the introduction of too many qualifying and saving clauses, one indeed is most important. Real animals, unlike imaginary test-tubes, are neither born mature, nor do they get on with the business of self-reproduction at once. There is always a pre-reproductive period during which animals are far from exempt from the hazards of mortality, and during this period the average reproductive value of an individual must therefore rise to a maximum, irrespective of whether or not it falls later. If my reasoning is correct—there is no time to go into details—the precession of the time of action of genes comes to a standstill at the epoch when the reproductive value is at a maximum, and it is *then* that senescence should be expected to begin. Professor R. A. Fisher[1] has pointed out that the actuarial prime of life of human beings and the age at which their reproductive value is at its maximum do in fact nearly coincide.

Even with such refinements as this, my proposals can hardly be said to add up to a self-sufficient theory. If we concede that

[1] In his *The Genetical Theory of Natural Selection*, reprinted by Dover Publications.

the force of natural selection is rather exactly weighted by the ages of the animals on which it operates, it is still far from easy to see in detail how senescence has become shaped into its distinctive pattern—the early onset and slow progressive fulfilment that the curve of the force of mortality so conspicuously reveals. Some of the agencies described seem to suggest a rather precipitous onset of senescence—more like that which befell the expatriates of Shangri-La than that suffered by the inhabitants of the world at large. But even allowing this shortcoming, I think it must be clear that the origin and evolution of senescence is not an insoluble *genetical* mystery, however mysterious it may be in other ways. The geneticist can see how it might well have happened; its occurrence does not outrage his sense of the fitness of things. So perhaps I was unduly disrespectful to Weismann's memory when I poked fun at his conjectures on senescence. In very broad outline they were probably not erroneous, at least in so far as natural selection was recognized as the instrument of its origin and perpetuation. I said earlier, as you may remember, that there was *some* truth amidst a good deal of what we can now see to be nonsense, and that it would stir up his successors to think up a more polished and cogent explanation. Not much more than this can be said of any biological theory of comparable pretensions, and I shall count myself lucky if I hear an equally sympathetic criticism of my own.

3

A Note on 'The Scientific Method'

It now seems to be agreed by those who direct our policy that the development and application of science is of immediate importance to England's economic welfare. So long as science could be thought of only as a means for the leisurely inception of an Age of Plenty, its benefactions could be postponed without fatal consequences. But we must now be satisfied with lowlier aspirations: science is to lead the state as the Red Queen led Alice—the most rapid progress is necessary with no higher ambition than to remain in approximately the same place as before. We now, therefore, hear a great deal about 'the scientific method', for the most part from people who might be quite upset if they were asked just what that method was supposed to be. The scholarly amateur might be heard to mumble something about the Question put to Nature and the *experimentum crucis*; the scientist speaks of quantitative method and the controlled experiment; the layman is often rude enough to think it no more than common sense. Let us press the question. How does scientific method differ from that used in other sorts of scholarly enquiry? What are the rules for making scientific theories? Just what does science *prove*? The answers to these questions have been quite widely agreed upon, but are not yet common property; they should be, and this essay is an attempt to make them so. Being no philosopher myself, it goes without saying that in what follows I claim proprietary rights only in what may be mistaken.

The currency of science consists of statements about 'matter of fact and existence'—*propositions*, they are often called, to distinguish them from questions, orders, outcries and suggestions, and some forms of the expression of abuse. But scientific knowledge is something more than the assembly of the facts reported by such statements: it has a corporate structure, a certain internal order and coherence of its own. There are several ways in which an order might be imposed upon them. For example, a man who wished to write a textbook about boron might begin by collecting under that heading all true statements made about it. This would not give the facts a peculiarly scientific structure, because the man who wrote the author's obituary notice would be expected, *mutatis mutandis*, to do very much the same for him. The grouping of statements by their subjects, objects, form or syntax, or the chronological order of the events recorded in them, though each has its special purpose, does not confer the structure of a *theory* upon them. Theories are sets of statements put into order by the relationship of entailing, and statements entailed are said to be 'explained' by those they follow from. Statements at the head end of the entailing are variously called premisses, axioms, postulates, or hypotheses—a luxuriant synonymy, since all are in effect, though not in form of origin, the same. Some attempt should be made to share out their legacy of meaning without spreading dissatisfaction equally among them all. We assert a postulate, and take an axiom for granted; hypotheses we merely venture to suggest. 'Premisses', when other people's, are usually so spoken of when not believed in. Scientists speak as a rule of their hypotheses. Some scientists seem to use the word 'theory' interchangeably with 'hypothesis', but this wastes a good word and should not be encouraged. A theory is the whole system of statements comprising hypotheses and the statements they entail.

All this is commonplace and rather uninspiring. What is of intense personal interest to many scientists is how an hypothesis ever comes to be devised at all. Its creation is evidently a leap upstream of the flow of deductive inference. One does not, as writers of detective stories seem to imagine, deduce hypotheses; quite the reverse, hypotheses are what we deduce things from. It was at one time thought that hypotheses could be arrived at by a rigorous logical process of 'induction', but even that humblest sort of hypothesis (for such it is), the simple collective generalization, defied these efforts to make it logically respectable, and it defies them no less resolutely to-day. Philosophers who now irritably contend that induction does not require their formal blessing forget that it was they themselves or their predecessors who first attempted the laying on of hands. Leaving aside those forms of scientific enquiry that may be purely documentary or descriptive in purpose—the determination of an atomic weight, say, or the anatomy of a mollusc—it seems that no attempt to solve a scientific problem can even be begun without the subsidy of some hypothesis, however dimly formulated or however vague. The first stage of textbook induction as I learnt it used to be the assembly of 'relevant' information; but what could it be relevant to, if not to the terms of some preconceived hypothesis? In my experience there is *no* stage in the working out of a scientific problem in which some hypothesis is not for the time being in office, and scientific activity comes instantly to a standstill without this sort of direction of its affairs. This, of course, says nothing about how hypotheses come into being. So far as I can tell from my own experience and from discussion with my colleagues, hypotheses are thought up and not thought out. One simply 'has an idea' and has it whole and suddenly, without a period of gestation in the conscious mind. The creation an of hypothesis is akin to, and just as obscure in origin as, any other creative act of mind. If science were an art we should call it inspiration, but as only astronomy has a Muse that will not do. Our leading philosopher[1] once called it 'a mere method of making plausible guesses'. The word *mere* rankles, for it is guesswork that must be imaginative, apt and technically informed. Too much learning may however be as dangerous as too little. All scientists know of colleagues whose minds are so well equipped with the means of refutation that no new idea has the temerity to seek admittance. Their contribution to science is accordingly very small.

It is right to point out, because of the irritating *mystique* that has grown up round it, that clinical diagnosis illustrates the act of hypothesis formation in an uncomplicated and fairly lucid way. The clinician seeks an hypothesis that will account for his patient's illness. There is no time in the course of his investigation during which *some* hypothesis is not in the background of his mind, and during its early stages there may be many. If his mind ends up blank after examination, that is not because no hypothesis sought admittance, but because all that did so had to be turned away. The experienced clinician is very well aware of the intuitive nature of the act of mind by which he hits on an hypothesis, but he sometimes fails to realize that this is the commonplace of scientific discovery: hence the fuss.

A scientific theory is propped up on either side, like Moses' arms before the Amalekites, by twin supports that together form its 'metatheory', and without these Reason cannot prevail. One part of metatheory, now called logical syntax, deals with the concepts of formal truth and falsity and the ordinances that govern the activity of deducing. Logical syntax is wholly the logician's business. The second part, semantics, more recent of origin and in lay circles now more fashionable, deals with the theory of the meanings of words and the ideas of material truth and falsity. The semantic problems of a science

[1] Bertrand Russell: *The Principles of Mathematics*, p. 11*n*, 1903.

have always been solved and are best solved by its own practitioners, and no more need be said about them here. But several things are worth saying about deduction. From the days of Sextus Empiricus onwards philosophers have confidently or more or less reluctantly affirmed that the process of deduction is simply the unravelling of tautology. Deduction renders explicit, discloses or makes manifest the information concealed within the axioms from which it issues; so far from adding new information, it merely attenuates it or makes it more dilute.* Thus the theorems of Euclid are but a few of the endless possible reaffirmations of his axioms; they exist as reproachful evidence of the mind's imperfection, because for a perfect mind the axioms would be enough. Deduction involves no creative act of mind and no imagination. The Mechanical Brain will one day undertake our deductive reasoning for us; to some extent it already does. The respect that our now queasy Frankensteins show for their intricate but guileless monster may be due to their realization that mathematics is but tautology after all. If that is so, it will serve them right for the qualms they have caused among the laity if such a Brain one day submits its candidature for the Wayneflete or Sadleirian Chair.

A second property of deduction is of the utmost importance for appreciating the validity of what is so often recklessly spoken of as 'proof'. The rigours of deduction are in one way curiously overrated: it proves to be quite a lenient discipline after all. For when it is said that one statement *follows from* another, deduction admits any combination between the truth or falsity of either except just one: that the first statement should be true and the second false. All that it guarantees among alternative possibilities is that what follows from a true premiss should be true. The dilemma of 'proof' is simply that an

*[H. A. Rowlands has put it admirably: deduction obeys a Law of Conservation of Knowledge.]

hypothesis may be false although the inferences drawn from it are themselves empirically true. This combination is by no means disallowed by logic. Consider that paradigm of empirical facts, the mortality of Socrates. If Socrates is a fish and all fish mortal, it follows with pitiless logical rigour that Socrates is mortal too. This is perfectly straightforward deduction, and Socrates' mortality is adequately so explained. How then does the scientist prove his hypotheses? The answer, now widely accepted, is that except in certain limiting instances he never does. No concept is so maltreated by lay usage as that of proof. In a strictly formal sense, accepted hypotheses remain perpetually on probation; one does not prove them true, though one may often act exactly as if they were. But what the scientist can often do with complete logical precision is to disprove hypotheses. If what follows from an hypothesis is false, then the hypothesis is false, and false in logic. This consequence of the asymmetry of the process of 'implying' is a central property of scientific method, and it influences experimental design in a direct and conspicuous way: many experimental designs are simply well-laid traps to lure on a so-called *null* hypothesis and then confound it. The precision of the act of disproof is thus very far from being a formalistic fancy. This does not mean, of course, that the accepted hypothesis is *merely* 'not disproven'; there are obviously degrees of certitude of conviction, but these are for the most part informally worked out. It is clear that an hypothesis gains in acceptability merely by its fitting in to a wider theoretical scheme of which it is a part. In this way hypotheses may sustain each other.* At all events, the scientist would soon be beggared by Descartes' first precept of intellectual enquiry—'*de ne recevoir jamais aucune chose pour vraie que je ne la connusse évidemment être telle*'. His own precept of

* [I confess here to the fault of having lumped all hypotheses together, as if they were a single logical species; for a careful analysis of their several forms, see *Probability and Induction*, by W. Kneale, Oxford, 1949.]

enquiry is the mirror image of this one, to accept nothing which is demonstrably false.

No hypothesis is admissible in science that accounts *only* for the facts it was expressly formulated to explain. Such an hypothesis is inadmissible not, as we have seen, because it cannot be verified but because it cannot even in principle be proved untrue. Nothing can be done with an hypothesis that has no 'extra-mural' implications, and its acceptance and rejection are equally acts of faith.

'Testing an hypothesis' is the act of examining these extra-mural implications. If they are true, the hypothesis is in some recognizable but obscure way strengthened; if they are false, the hypothesis is false. All fish have gills, but Socrates proved to lack them; we must therefore think up some other explanation of his having died. In practice, of course, we are not often lucky enough to deal with such crisp disjunctions. An hypothesis is less often outright false than merely inadequate, and not beyond the help of running repairs.

The act performed to test an hypothesis may be called an 'experiment'. It is best to use the term in this simple and clean-cut way, rather than to follow common use in restricting its terms of reference to some sort of active messing-about with nature. A 'mere observation' may in this sense be an experiment, and if activity is insisted upon as a criterion, it may be answered that even the merest observations cannot be made from a supine position. The hypothesis which predicted the existence of a planet Neptune was tested by the experiment of directing a telescope towards a certain predicted region of the sky. The existence of an experimental method in this generalized sense is what distinguishes the scientific method from that of any other sort of scholarly enquiry, and it is to this method that science owes its power. It will be noticed that I have said nothing here about the virtues of metrical analysis or the controlled experiment, or all the many other things that to the layman seem to be so characteristically scientific. These things belong not to scientific method in its more formal sense but to the theory of experimental design and scientific analysis, and the exacting requirements of scientific *reportage*. Obviously an experiment must be done in such a way as to give an unambiguous answer, and in the examination of events one must aspire to put on record that which is indeed the case. All this belongs to the technology of scientific method.

The growth of science is organic and not accretionary. The structure of knowledge built up by the prosecution of the scientific method as I have outlined it is a tapering hierarchy of hypotheses, the more general counting the less general among their consequences, the least general—the ordinary colligative inductions—finally touching down in a multitude of particular statements about fact. The structure of scientific knowledge is therefore in the outcome logico-deductive, and this is the form in which what Berkeley called the Grammar of Nature is finally written down. It is already a record of some grandeur, though the greater part has yet to be compiled.

4

A Commentary on Lamarckism

1. INTRODUCTION

I begin by excusing myself the task of making any detailed exposition of the evolutionary teachings of the Chevalier de Lamarck. Darwinism, we know, is Wallace's word, and Lamarckism is not Lamarck's; and although the words stand for doctrines which their eponymous authors would have no difficulty in recognizing as their brain-children, it is their latter-day growth and present stature that must occupy the whole of our attention. Nor will it be profitable to carry out a semantic autopsy upon expressions like 'the survival of the fittest' and 'the inheritance of acquired characters'. 'Fitness' is now so defined as to make Spencer's phrase a tautology, and its use tends to perpetuate the mistaken belief that the famous Malthusian syllogism is a necessary part of the logical structure of Darwinism (see Introduction). As to the 'inheritance of acquired characters', its last solemn rites have been capably intoned by Woodger (1952), and there can be no case for having it disinterred.

The purpose of this introductory section is (*a*) to present in the simplest possible terms the essential difference between Darwinian and Lamarckian interpretations of the hereditary process, and (*b*) to show that inheritance that may be represented as Darwinian on one plane of analysis may be represented as Lamarckian on another.

Consider for this purpose a population of streptococci

(though other micro-organisms will also serve). Streptococcal infections are usually treated by the administration of sulphonamide drugs or of antibiotics of fungal or bacterial origin, such as penicillin or streptomycin. It is a common observation of clinical practice (and one which can be reproduced by experiments *in vitro*) that the prolonged exposure of a population of streptococci to penicillin, at concentrations which fall short of bringing about its complete destruction, *may* lead to the evolution of a 'resistant strain'. i.e. a population which can flourish unchecked at a concentration of penicillin that strongly inhibited the growth of the parental organisms. The transformation is heritable, for resistance once acquired long outlives the stimulus that originally called it forth.

By disregarding all subtleties of interpretation, the 'training' process may be represented in alternative ways. The first is illustrated by Fig. 4; the shaded circles represent resistant

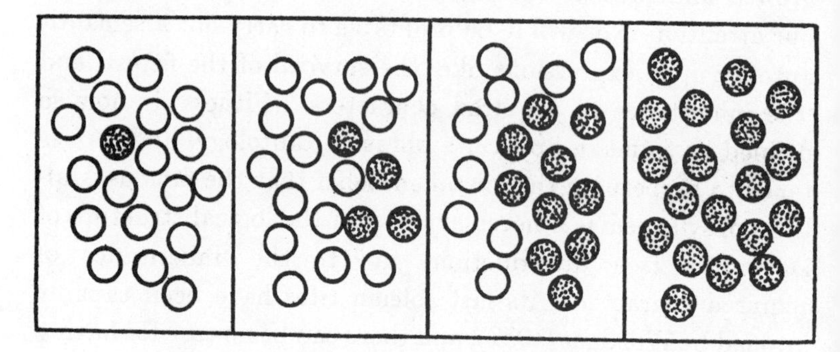

Fig. 4. The development of penicillin-resistance in bacteria, according to a Darwinian interpretation. For explanation see text.

forms. It is presumed that the original bacterial population was heterogeneous and contained genetic variants endowed with a relatively high degree of resistance to the action of penicillin. In the normal course of events—in a penicillin-free microcosm—these variants stand at no selective advantage; but under the influence of penicillin they proliferate more rapidly than their unresistant neighbours and so eventually become the prevailing forms.

The second is illustrated by Fig. 5. It is presumed that the enzymic organization responsible for the metabolic activity of

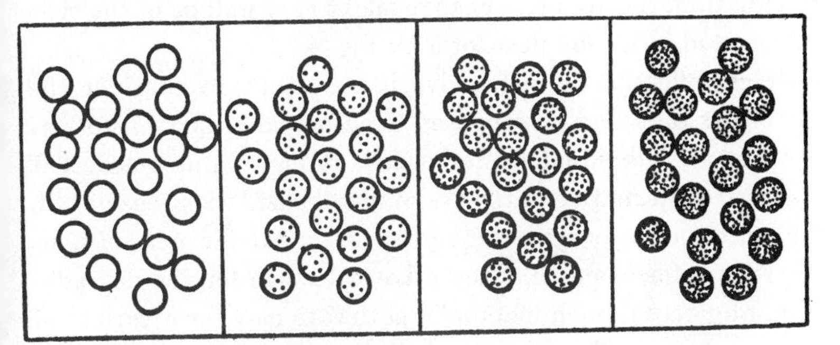

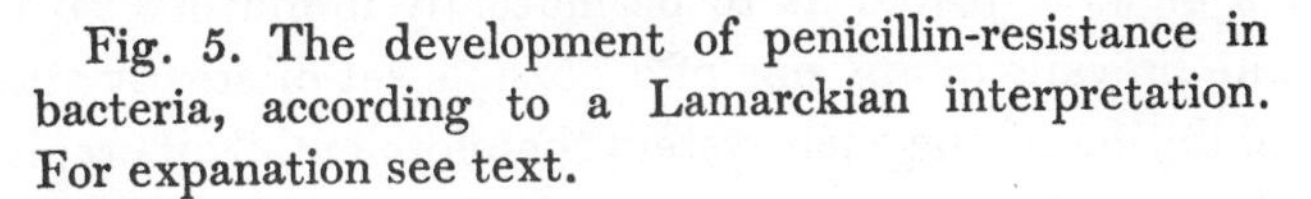

Fig. 5. The development of penicillin-resistance in bacteria, according to a Lamarckian interpretation. For expanation see text.

each individual is progressively altered, as the shading indicates, and that the change so produced is heritable. The change in the properties of the population is therefore the sum of the changes brought about within each individual. The difference between the two interpretations, Darwinian and Lamarckian, is that the one presents adaptation as a change in the genetical structure of a *population*, and the other as a change in the genetical structure of an *individual.* These are avowedly extremes, for they are in no sense mutually exclusive. On the contrary, any inherited difference in the 'Lamarckian' adapt ability of individuals must of necessity become the subject of selective discrimination.

So much is commonplace. Let us now consider not a single population but an assembly of such populations, supposing each one to be anatomically separate and distinct. The entire assembly of populations is now subjected to training by penicillin, and it is found that each individual member becomes progressively adapted to resist its action. In a world in which such populations were the analytical units, such a transformation would be called 'Lamarckian' in whatever sense the scheme illustrated by Fig. 5 may be so described. But *within* each population, the adaptive change might very well be of the type illustrated in its simplest form by Fig. 4.

This reflection is instructive if we return to consider the activities that may be supposed to accompany the transformation of an individual bacterial cell. It may be assumed that there are alternative pathways of metabolism within each cell, i.e. alternative enzyme sequences or metabolic gearings, as there are, for example, alternative pathways for the degradation of glucose. Such metabolic pathways may for a variety of reasons be so adjusted as to be mutually inhibitory, so that only one prevails in any one of a possible set of steady states. The inhibition of one such system therefore entails its replacement by another. In other words, as Hinshelwood (1946) has pointed out, the Lamarckian transformation illustrated by Fig. 5 may be Darwinian at the lower analytical level represented by the enzymic population or complex of intersecting metabolic pathways within the individual bacterial cell. Such a description would be pointless for any except explanatory purposes, but it shows that no discussion of the rival interpretative powers of Darwinism and Lamarckism can have any useful outcome unless a certain analytical level is defined and adhered to. Hereafter we shall be concerned with individual organisms as analytical units, for it is only in this context that the rivalry is of any moment.

The case for and against Lamarckism may be set out for analysis in a variety of ways. Guided by the reflections of Baldwin and Lloyd Morgan, I shall present it first in what philosophers would call a 'weak' or general form, and then in a 'strong' or special form. The weak form may be so described because it merely proposes the existence of a certain mode of origin of inherited differences, without expressing any opinion about the actual mechanism by which those differences have come into being; but they are differences which, unlike so many, are *open* to a Lamarckian interpretation of their origin. The 'strong' form goes farther and positively affirms that the Lamarckian interpretation left open by the weak formulation is in fact the correct one.

2. THE 'WEAK' FORM OF LAMARCKISM

The 'weak' form, then, may be expressed in these terms:

> Modifications acquired in each member of a succession of individual lifetimes, as a result of recurrent responses to environmental stimuli, may eventually make their appearance in ontogeny even when the environmental stimuli are absent or are deliberately withheld. . . .

We may proceed at once to strengthen this formulation by making it in one respect a little more particular:

> . . . and the age of appearance of these modifications in ontogeny will eventually anticipate the age at which environmental stimuli could in any case have been responsible for them

This clause is separated from the main body of the formulation merely to emphasize the fact that it *is* formally separable, but we shall adopt the fuller and more particular formulation for the good reason that every example we shall consider will be shown to satisfy it. It must again be emphasized that the 'weak' formulation neither embodies nor presupposes any hypothesis about how acquired character differences become inherited character differences: it merely states that they become so.

Before proceeding to the discussion of special examples, we may ask: of which character differences may it plausibly be argued that they have arisen in the way that has just been proposed? The answer is a very simple and obvious one: they are character differences having the distinctive property that, although they are in fact 'laid on' by development, *they could in any event have been fashioned in an individual's own lifetime merely as a response to differences of use.*

Consider, for example, the difference between the characteristically thick, richly stratified and mitotically active epidermis on the sole or heel of the foot and the thinner and more delicate epidermis that covers the greater part of the rest of the body. The difference is at least in large part of purely developmental origin, i.e. of the same sort as that which distinguishes epidermal cells from pancreatic or thyroid cells. It does *not* arise, as we are at first tempted to think, because of the chronic chafing and general mechanical stress that soles of feet are obliged to put up with (although such stimuli can certainly exaggerate the difference). Both the human being and the guinea-pig are born with a thicker epidermis on the sole of the foot than elsewhere on the body.* Such a difference is therefore developmentally prefabricated; it could not have arisen as an adaptive response *in utero* because the foetus treads water in so far as it treads at all.

The argument may be reinforced by experimental proof. If the difference between trunk and sole-of-foot epidermis arose merely because the latter is habitually trodden upon and otherwise abused, while the former is not, then sole-of-foot epidermis should revert to the condition of relatively delicate and quiescent body skin after transplantation to a protected position elsewhere on the body. Billingham and I (1948, *a*, *b*), have done this experiment on the guinea-pig, and find that sole-of-foot skin conserves its distinctive thickness, stratification and mitotic activity even two years after its transplantation to a

* [A fact well known to Darwin, and commented upon by others since.]

completely protected position on the ordinary skin of the chest.

So much for the evidence that the difference between sole and body epithelium is of developmental origin, i.e. is an inherited difference between the somatic cells that arise by fission of the zygote. It must now be shown that even if the difference were *not* of developmental origin, it would be almost exactly reproduced within an individual's own lifetime as a response to differences in the habit of use; it must be shown that if guinea-pigs or human beings were in fact born with a thin and delicate epithelium on the soles of their feet, ordinary use would soon toughen and thicken it.

There can be no reasonable doubt that this would be so, because a normally quiescent epidermal epithelium can easily be induced to thicken in response to chronic irritation. Corns are so formed on the thin skin of the dorsum of the toes; callosities on the hands develop as a response to chronic chafing. A corn has a histological structure very closely similar to that of the skin on the heel of the foot, with a deep, stratified, vigorously dividing epidermis, a thick pad of compact cuticle, and tall, steeply rising dermal papillae. The difference is that corns and callosities do not last much longer than the mechanical stimuli that provoked their formation: corns subside with the wearing of shoes that fit; callosities may be cured, as they may also be avoided, by wearing gloves. Evidently the epidermis has the *capacity* to thicken in response to mechanical abuse. In the epidermis of the sole or heel, this thickening is developmentally anticipated and does not depend for its maintenance upon the continued stress of use; and yet, were it not so anticipated, stress of use could be relied upon to reproduce it faithfully. *All* adaptations that are open to a Lamarckian interpretation have this distinctive character: that they represent differences of developmental origin that can be faithfully mimicked within an individual's own lifetime by differences in mode of use.

Contrast the state of affairs that has just been described with another difference between the races that together constitute the epidermal (or ectodermal) epithelia: the difference between 'ordinary' body skin and the compact, non-flaking, perfectly transparent epithelium of the cornea. Here again, the difference is of developmental origin; nor is it kept in being by the fact that the corneal epithelium lives in an environment very different from that of ordinary skin. The cornea is non-vascular, moist and cool; ordinary body skin is vascular, dry and (being dry) warmer than the cornea. Yet if corneal epithelium is transplanted to an area formerly occupied by ordinary body skin, and *vice versa*, the distinctive differences between the two remain (Billingham and Medawar, 1950). The property that distinguishes this case from the one just considered is this: that if the difference between corneal and ordinary body-skin epithelium were not of developmental origin, it could *not* be reproduced within an individual's own lifetime by difference of environment or of mode of use. The difference will be established by developmental mechanisms, i.e. by the appropriate segregations within the lineage of cells arising by division of the zygote, or not at all.

Let us call the difference between corneal and body-skin epithelium a difference of Class A, and that between sole-of-foot and body-skin epithelium a difference of Class B. To these should be added a third category of difference, of Class C (Abercrombie, 1952): one which is not developmentally prefabricated, but which may arise purely from difference of environment or of use.* The pigmentary cells (melanocytes) of the epidermis of the two sides of the face or the two arms may

* [This rather arid terminology was based upon that of an article published in *New Biology*, **11**, p. 10, 1951. Much better, because self-explanatory, is C. H. Waddington's (*loc. cit.*, 1953): Class C adaptations are 'exogenous', Class A 'endogenous', and Class B 'pseudo-exogenous'. I was not able to benefit from these suggestions, because the present article was two years 'in the press'.]

be supposed to have the same properties and to be present in the same numbers. Expose one side of the face or one arm to sunlight, and it will become darker than the other, no matter why. The difference of degree of pigmentation is caused by differences of environmental stimuli, and by them alone.

Differences of Class B, those which are open to a Lamarckian interpretation of their origin, are commoner than is usually supposed. The flexure lines of the palm of the hand provide a splendid example. The bolder flexure lines are easily visible in the twelve weeks' foetus, and even if the foetus is not incapable of clenching its hands, it would be idle to suppose that flexure lines were formed by the imprint of habitual use. Use neither forms them nor keeps them in being, for the plastic surgeon tells us that if skin grooved by a flexure line is displaced or transplanted to positions in which it is not normally creased or folded, the flexure lines will nevertheless persist. Yet ectopic flexure lines *can* be formed by habitual creasing of the skin—by frowning for example, or raising the eyebrows; and we can therefore be quite confident in saying that if the palmar flexure lines were not developmentally prefabricated, a very exact copy of them would soon be formed in the ordinary run of everyday use. Ectopic flexure lines are exactly analogous to the corns and callosities that were called in evidence in our earlier example. They differ from the 'inborn' flexure lines because they disappear with the withdrawal of the stimulus that was responsible for their formation.

In saying that flexure lines and thickened soles are of developmental origin, I do not wish to deny that use within an individual's own lifetime may not make flexure lines bolder and soles thicker still. Why should it not be so, if folding and chafing of the skin can cause the formation of ectopic flexure lines or epidermal thickenings elsewhere on the body? It is likely, but not certain, that the ordinary use of a joint or bone completes the otherwise purely developmental differentiation of articular surfaces and the patterns of bony trabeculae. With this qualification, the mode of development of joints is closely comparable to that of flexure lines and thickened soles. We are born with working joints, cartilage lined, encapsulated, lubricated with synovial fluid, and with their apposed surfaces having just that complementarity of structure which might be expected to arise from the mechanical exactions of ordinary use. Foetal movements have only a small part to play in fashioning the final structure, for joints develop from primordia cultivated *in vitro* or transplanted to the chorio-allantoic membrane—positions where no movement can occur. In spite of that, functional cartilage-lined and encapsulated ectopic joints *can* be formed in an individual's later lifetime if by accident (or orthopaedic artifice) two mobile bony surfaces are apposed to each other, as in an unhealed fracture. Here too then, it appears, the mechanisms of morphogenesis exist in duplicate, and what could be formed by use is in fact formed by pre-emptive differentiation.

In the foregoing account I have deliberately confined myself to familiar everyday examples of pre-emptive differentiation in metazoa. (Micro-organisms come later.) The more esoteric examples collated by Wood Jones (1943) [1] appear to me to introduce no distinction of principle, and an explanation valid for the one set should be valid for the other. Each represents a character difference of developmental origin that could also have arisen as a direct adaptive response to difference of use within an individual's own lifetime.

All such adaptations are open to a Lamarckian interpretation of their origin. All that remains to establish a strong *prima facie* case is evidence that acquired character differences can

[1] For example the squatting facets between tibia and ankle-bone in Panjabi (but see Medawar, 1952) and the callosities on the 'knees' of the African wart-hog; to which add Kukenthal's strange story of the dugong's teeth, as it has been recounted by de Beer (1951).

become inherited character differences under conditions that formally exclude the action of natural selection. It will be clear from Section 3 that evidence of the occurrence of any such transformation in metazoa is still wanting. We must conclude that although what I have called 'Class B' adaptations *might*, unlike so many others, have arisen in Lamarckian fashion, there is no unambiguous evidence that they have done so.

This answer is quite widely thought by laymen and ill-informed zoologists to be shifty-eyed and evasive, and the reason is not far to seek. It is believed, quite mistakenly, that eligibility for a Lamarckian interpretation is in some way discreditable to Darwinism. The truth is quite otherwise. The developmental pre-emption of what would otherwise be acquired character differences is, with most adaptations of Class B, of conspicuous selective advantage. If it is an advantage to have thickened soles at all, it will be particularly advantageous to have them ready-made—ready for use the first time the foot touches the ground. And what could be more biologically inept than a state of affairs in which the several joints, only roughly fashioned at birth, had to be 'run in' to complete their particular articulation patterns during the lifetime of each individual? In so far as the plausibility of a Darwinian argument turns upon the demonstration of conspicuous selective advantages, Class B adaptations are as amenable to Darwinian explanation as any other. It is indeed an explanation with many obscurities and shortcomings; but all I seek to emphasize is that none of them is peculiar to adaptations of Class B, i.e. peculiar to adaptations of the only kind for which a Lamarckian explanation is theoretically admissible. The adaptive value or 'selective advantage' of having developmentally prefabricated flexure lines is far from obvious; but so also is, for example, the adaptive value of many of the antigenic variants that determine blood-group polymorphism and the incompatibilities revealed by grafting—differences upon which a Lamarckian interpretation has no bearing whatsoever. Flexure lines are mysterious, but not mysterious in any way that is particularly discreditable to Darwinism.

This section may well conclude with a description of an important experiment in which Waddington (1952) has demonstrated the genetical pre-emption of a change originally brought about by environmental means. If fruit flies are subjected to a mild temperature shock shortly after pupation, a certain proportion develop without the cross veins that bridge the principal veins of the wings. Flies of this susceptible fraction were bred from, and their offspring again shocked; the susceptible fraction again bred from, and so on. The proportion of susceptible flies steadily increased, as was to be expected; but from the twelfth generation onwards, the cross-veinless condition began to appear in flies which had received no temperature shock at all. Selection has thus, in effect, converted an acquired into an inherited character difference.*

The gist of the foregoing argument is as follows. Darwinism and Lamarckism may be thought of as competing interpretations of the origin of inherited character differences in metazoan individuals. An examination of these character differences shows that only a certain category, described as Class B, is open to a Lamarckian interpretation at all. But there is, on the one hand, no evidence to suggest that the Lamarckian interpretation is the correct one; and, on the other hand, Darwinism is no less competent to explain the origin of Class B adaptations than the origin of any other.

3. THE 'STRONG' FORM OF LAMARCKISM

The weak form of Lamarckism, which we have seen to be unobjectionable, is purely descriptive in intent; it merely

* [Some of Waddington's more recent experiments are reported in *Evolution*, **10**, p. 1, 1956.]

describes a biological history of the origin of certain inherited character differences. The 'strong' form of Lamarckism is the weak form strengthened (in the sense of being made more particular) by the categorical statement that the origin of acquired character differences is accompanied by the origin of *adaptive genetical* differences in the individuals in which they are induced. By an 'adaptive' genetical change is only meant such a change as will reproduce the character difference originally elicited by the environment: the enlargement of a particular muscle by habitual use must be accompanied by such a genetical change as will entail the enlargement of that muscle. The qualification 'adaptive' is therefore of central importance. That differences of environment or of 'treatment' may bring about genetical transformations has not been in dispute since Muller's demonstration, now a quarter of a century old, of the mutagenic action of X-rays, and the number of physical and chemical treatments known to increase mutation rate is being steadily added to.

Lamarckists do not suppose that adaptive genetical changes are completed within a single generation; the 'strong' form of Lamarckism may therefore be expressed in such a way as to take this qualification into account:

> The repeated induction of character-differences within the lifetimes of individuals of successive generations is accompanied by a genetic change in each individual, the change being such as eventually to reproduce the character-difference elicited by environmental stimuli even when those stimuli are withheld.

It will be clear that the only acceptable evidence for Lamarckian inheritance in the strong sense will be that in which the possibility of selection is scrupulously eliminated. This section will begin by a consideration of four examples of supposedly Lamarckian inheritance in higher animals, choosing the experiments on the grounds that they have been conducted with care and reported in sufficient detail to make an appraisal possible, and avoiding those in which there is a suspicion of corrupt advocacy.

3A. LAMARCKIAN INHERITANCE IN HIGHER ORGANISMS

(i) *The inheritance of eye-defects induced by specific antisera.* Guyer and Smith (1918, 1920, 1924), although not themselves 'particularly interested in establishing or disestablishing any ism', claimed to have shown that eye defects induced in rabbit foetuses by the injection of pregnant does with anti-lens serum were reproduced in successive generations born of the affected rabbits. In a representative experiment, rabbits' lenses were pulped and injected into chickens to elicit the formation of anti-lens precipitating antibodies. The antiserum so formed was injected into pregnant does. A small proportion of the offspring were born with eye abnormalities ranging from opacity and mis-shapenness of the lens to an apparently complete 'liquefaction'. These induced differences of eye structure were inherited, in roughly the manner of a Mendelian recessive, through both male and female lines.

With variations that may have been significant, these claims were tested by three independent groups of workers (Finlay, 1924; Huxley and Carr-Saunders, 1924; Ibsen and Bushnell, 1931, 1934) with negative results. The findings of Guyer and Smith therefore remained in the penumbra of unexplained anomalies until Sturtevant (1944) proposed a *prima facie* genetical case for their acceptance. Following a train of thought started by M. R. Irwin and J. B. S. Haldane he argued that, in as much as there is in general a one-to-one correspondence between particular antigens and particular genes, an antigen may be 'a rather direct gene product' and may be imprinted with some of the structural specificity of the gene. 'If a particular gene is responsible for the formation of a given antigen, there is a possibility that antibodies induced by this antigen may react with the gene.' In other words, the anti-lens serum, in addition to acting directly upon the foetal lens, may have altered in a genetically reproducible way the structural specificity of one or more 'lens genes'. Sturtevant refers to unpublished (and apparently still unpublished) evidence of R. R. Hyde in support of the original authors' claims.

Guyer and Smith's experiments are plausible in a purely immunological sense, quite apart from the fact that they were done in a period when the authors could hardly have hoped for a genetical benediction. An 'anti-kidney' or 'anti-mesenchyme' immune serum would be expected to be quite ineffective, because the immune bodies would be promptly absorbed by the corresponding maternal tissues and so denied access to the foetus. But the lens of adult rabbits is avascular; anti-lens antibodies should not therefore be absorbed by the mother but should be left free to act upon the vascularized lens of the foetus. Nor is there any doubt that antibodies can reach the rabbit foetus—not through the placenta, as was formerly believed, but through the yolk sac (see Brambell, Hemmings and Henderson, 1951). Unfortunately, there is discrimination against antibodies ('heterologous antibodies') formed in an organism of a foreign species, and this, combined with the very decided toxicity of foreign serum as such, makes one regret that Guyer and Smith did not persevere with the experiments in which they tried to elicit anti-lens antibodies from the rabbit itself.

We must not, however, be led astray by speculations on whether or not the phenomena described by Guyer and Smith *could* happen; the problem is whether or not they *do* happen, and the answer to this problem is at present open.

(ii) *The inheritance of learned behaviour differences in rats.* McDougall argued that a fair test of Lamarckian inheritance should be one in which the acquired character difference represented the outcome of an active and (in the everyday sense) 'purposive' response by the subject, and should be such that the results were open to quantitative assessment. He therefore studied the inheritance of the acquired ability of rats to learn one of alternative methods of getting out of a water trap.

The trap was a water bath with a central entrance ramp and two exit ramps, one brightly illuminated and so wired as to give a tetanizing shock, the other dim but not electrified. The two exits were alternated to prevent the complication of the experiments by the learning of left-handed or right-handed habits of emergence. The rats came from the inbred stock of the Wistar Institute, and were divided into three groups of which two were bred from at random, or at least without avoidable selection. The three groups were (*a*) untrained controls; (*b*) experimental rats that had been trained in the tank; and (*c*) rats which had been through the tank tests but which, instead of being bred from at random, were deliberately selected for breeding from those which showed the worst performances. The criterion of learning status was the number of tests that had to be given before an individual scored twelve correct choices of exit successively.

The results of McDougall's experiments, reported over a period of years in the *British Journal of Psychology* (1927, 1930, 1938; McDougall and Rhine, 1934), were as follows. The tank-trained rats of group (*b*) improved in performance from a score of 120 errors in the first generation to only 36 in the thirty-fourth generation of non-selective inbreeding. Unfortunately, the rats 'negatively' selected from the dullards of each generation (group *c*) improved from performance scores of 215 to 43 over the same period, and the untrained controls improved from 149 to 102 over a period of only four years.

Careful independent repetitions of McDougall's work by Crew (1936) and Agar, Drummond and Tiegs (1935, 1948) failed altogether to confirm his empirical findings; they were not scrupulously exact repetitions, it is true, but embodied refinements that increased the precision of the experiments without in any way affecting the principle of their design.* McDougall's results are therefore on a somewhat different footing from those of Guyer and Smith.[1] What part could selection have played? In theory no part, for the experimental subjects had been inbred for a sufficient number of generations to justify the prevailing theoretical assumption that they were genetically uniform and homozygous. In practice, this presumption seems to have been unduly optimistic: Loeb's work (1945) on the transplantation of tissues between members of the highly inbred Wistar strain of rats revealed incompatibilities that can only have been due to flagrant heterozygosity. Guinea-pigs and mice, by contrast, become completely tolerant of grafts transplanted between members of an inbred line after a much less prolonged regimen of inbreeding. McDougall's stock may, then, have been more heterogeneous than is usually supposed—and, as Drew (1939) has made clear in his admirably succinct review, there is plenty of evidence that differences of intelligence in rats, as measured by maze performances, are perfectly amenable to selection.

It may of course be argued that McDougall's adverse selec-

[1] Haldane (1951) makes the comment that McDougall's colleague and pupil, Rhine, was conducting experiments in paranormal cognition in the same laboratory, and points out the inconsistency of presenting evidence in favour of paranormal cognition in human beings without taking into account its effect on the outcome of such experiments as McDougall's.

* [The final report on this long and important experiment has now been published: W. E. Agar, F. H. Drummond, O. W. Tiegs and M. M. Gunson, *Journal of Experimental Biology*, **31**, p. 307, 1954.]

tion experiments prove that his results could *not* have been due to inadvertent selection. Unfortunately, the results from group (*c*) raise the new difficulty that improvement was more striking in the line perpetuated by dullards than in the unselected experimental stock; and there appears to have been a general secular improvement in the group (group *a*) which had not been exposed to the tank tests at all. McDougall's case must stand or fall by the empirical results, and these have not been confirmed.

(iii) *Melanism in moths*. 'The spread in industrial districts of melanic forms of Lepidoptera is . . . one of the most considerable evolutionary changes that has ever actually been witnessed' (Ford 1940). The change is widespread and has been rapid.

It was argued by Heslop Harrison (1926, 1928) that melanism is an induced and inheritable adaptive change: food plants in industrial areas were held to be contaminated by metallic fumes, and Harrison claimed to have induced the formation of melanic mutants by feeding larvae of the moth *Selenia bilunaria* on hawthorn leaves which had absorbed small quantities of salts of manganese and lead. Hughes (1933; cf. also Thomsen and Lemche, 1933) repeated Harrison's experiments with six generations comprising 3265 individual moths and found no melanic forms among the treated or the untreated; he adds that manganese salts are present in normal plants and are not present to excess in plants of industrial areas.

There is a clear-cut alternative explanation of the spread of melanism in moths, for which we are indebted to Ford. Melanism is a mutant of regular occurrence in many species of Lepidoptera from non-industrial areas; there is therefore a clear case for supposing that such mutants have been selected for their superior viability in the smoke-stained countryside of industrial districts. Indeed, the experience of many workers has been that certain melanic mutants are tougher and more viable than the ordinary paler forms; presumably they have failed to spread to non-industrial areas because their advantage in toughness is more than outweighed by their greater conspicuousness.

(iv) *Sladden's experiments on the inheritance of altered food habits in stick-insects.* These are perhaps the best of the experiments that purport to demonstrate Lamarckian inheritance; all sorts of genetical complications are avoided by the fact that reproduction in the subject species is parthenogenetic. Sladden (1934, 1935; Sladden and Hewer, 1938) studied the inheritance of the acquired ability of stick-insects of the species *Dixippus morosus* to subsist upon ivy instead of their normal diet, privet. The life cycle in this species is 9-10 months long, and somewhat more than 500 eggs are produced by each individual.

The insects feed at night, and must feed every night. The alternative foods were offered for consumption in such a way as to provide a reliable measure of their degree of acceptability. In the 'presentation test', ivy and privet were offered on alternate nights, the privet being necessary to keep the insects alive if they failed to eat sufficient ivy. Acceptability was measured by the number of trials necessary before the final acceptance of ivy. In the 'preference test' ivy and privet were thrice offered simultaneously: the result was scored as 'ivy preference' if ivy was chosen on all three occasions, and so for privet; otherwise the result was held to be nonindicative.

After six generations there was a clear-cut increase in the acceptability of ivy, but it is noteworthy that a high proportion of this increase occurred in the first generation after the first presentation of ivy. There is also an echo of the difficulties that bedevil the interpretation of McDougall's work, in that the control insects, reared upon privet throughout, also showed a distinct increase in preference for ivy. Sladden's own interpretation of this finding, which turns upon seasonal changes of food preference, is unconvincing.

These are good experiments: the facts are well set out and

their truth is not in question. Thorpe (1938, 1939) has however suggested an alternative and rather unexpected interpretation based upon the fact that insects are susceptible of a high degree of olfactory conditioning, in the sense that odours normally distasteful to adults may be acceptable if larvae are exposed to them early enough. For example: the ichneumon fly *Nemeritis canescens* normally lays its eggs in the Mediterranean flour moth *Ephestia kuhniella*, and is strongly attracted by the smell of its normal host. It does not normally lay eggs in, and is not normally attracted by, the smell of the related wax moth *Meliphora*. But if the ichneumon flies have been deliberately reared in *Meliphora*, or have been exposed to its larvae shortly after emergence from the cocoon, then they *do* show a strong attraction to *Meliphora*. This transformation of host preference was complete in one generation; ten successive generations of rearing on *Meliphora* did not increase it.

With this and other evidence of similar import in mind, Thorpe therefore suggests that, in Sladden's experiments, some olfactory emanation from ivy caused a conditioning which increased its acceptability to stick-insects. Enough might arise from the egg to condition the newly hatched nymphs, particularly if their first food is egg-shell. This does not account for the progressive increase in the acceptability of ivy over six generations, but as the greater part of this increase occurred after the first generation, and as there was some increase in tolerance by the controls, it is difficult to regard this as a grave shortcoming of Thorpe's explanation.

These four examples inspire one with no confidence in the applicability of the Lamarckian scheme of inheritance to higher animals. Two are susceptible of clear-cut alternative explanations; a third, McDougall's, is open to question on the grounds of empirical fact; and the fourth, that of the inheritance of induced eye defects, is urgently in need of reinvestigation. I am not aware of any experiments that have a greater claim upon our attention than these four, though of many which have less. It is therefore the generally held view that the case for Lamarckian inheritance in metazoa is unproven.

3B. LAMARCKIAN INHERITANCE IN MICRO-ORGANISMS

A mode of inheritance which satisfies the definition with which this section began is demonstrated by non-cellular organisms. In such organisms the entire body substance participates in the act of reproduction, so that the argument against Lamarckism which turns on the physical inaccessibility of the germ plasm to environmental influences loses much of its force.

Two examples will be cited.

(i) *The inheritance of acquired resistance to antisera in Paramecium aurelia.* Paramecia may be immobilized or killed by the incorporation into their culture-media of an antiserum formed by injecting suspensions of whole individuals into rabbits. If Paramecia are cultivated in sub-lethal concentrations of antiserum, their progeny acquire a resistance to its action under conditions which (it is now known) completely exclude the mere selection of the more resistant forms for propagation. Resistance so acquired is retained for many generations of asexual fission—in some varieties, through sexual fission as well—in the complete absence of the stimulus which originally brought about the transformation. Evidently the antiserum has initiated a heritable change.

These phenomena have been studied in recent years by Bernheimer and Harrison (1940, 1941), Harrison and Fowler (1945, 1946) and Kimball (1947); most of our information, however, derives from the detailed and systematic genetical analyses of Sonneborn (reviews 1949, 1950) and more recently of Beale (1952).*

In very brief outline the evidence may be summarized thus. An individual *Paramecium aurelia* belongs to a variety—essentially a species; to a 'type', which is an assembly defined by mating compatibilities and so equivalent to a sex; and to a stock. A stock is the progeny of a single homozygous individual. Within a stock, an individual may display one and (except while a tranformation is actually afoot) only one of a distinct set of surface antigens defined and labelled by their power to elicit specific antibodies from the rabbit. Differences of antigenic composition between the individuals of a stock are heritable, but they depend upon differences of cytoplasm and not upon differences of nuclear genes. Different *stocks* are distinguished by different combinations of the antigenic characters that may be displayed by their constituent members, and these differences of antigenic potential are governed by differences of nuclear genes. (It seems likely that the same gene loci are represented in all the stocks of a given variety, and that differences of antigenic composition between stocks depend upon different representations of the alleles of these loci.) *Within* a given stock, however, it is an inherited cytoplasmic difference that discriminates between the range of antigenic possibilities governed by the prevailing nuclear constitution.

If an individual or an assembly of similar individuals is exposed to a sublethal concentration of the antibody directed against the prevailing surface antigen, a heritable transformation is brought about, in consequence of which the prevailing antigen is replaced by another member of the set characteristic of the stock. The effect of the transformation is to confer resistance to an antibody upon the progeny of an individual which was formerly susceptible to it. It is of some importance

* [The most comprehensive modern summary of this work is *The Genetics of Paramecium aurelia*, by G. H. Beale, Cambridge, 1954.]

that such transformations may also be brought about, though (so far as present knowledge goes) more slowly, by a variety of 'non-specific' stimuli such as changes of temperature or nutritional status, or by treatment with enzymes (Kimball, 1947).

It is clear that the cytoplasm of Paramecia is malleable in a way completely foreign to our conception of the propagation system of the chromosomes, and that this malleability endows them with what is, in effect, a cytoplasmic genetic memory. We shall not delay with interpretations of the mechanism of the adaptive response, except to say that all turn upon the idea of an intracellular competition, whether between self-perpetuating cytoplasmic particles or between reaction sequences that are mutually inhibitory and so mutually exclusive. Such an interpretation gives point to Hinshelwood's comment that inheritance which is Lamarckian in terms of cells should be described as Darwinian at the level of cellular ingredients.

(ii) *Adaptive transformations in micro-organisms.* The 'training' of micro-organisms is an old story in bacteriology, but it is only in quite recent years that it has been seen to have an educational import for zoologists as well as for bacteria.

Bacteria may be trained to use lactose or glycerol instead of glucose as a source of carbon; nitrates instead of atmospheric oxygen; ammonium salts instead of amino-acids as a source of nitrogen; and so on. They may also be trained to resist antibiotics and other growth inhibitory agents to which they were at first susceptible.

The interpretation of the mechanism of these changes is complicated by two facts: (*a*) bacteria are too small for it to be possible to study their individual histories in sufficient detail, so that the behaviour of individuals must be inferred from the behaviour of bacterial populations; (*b*) the gene is not known as a unit of segregation but only as a unit of mutation. The modern analysis of recombination phenomena in viruses and bacteria (Delbrück and Bailey, 1946; Tatum and Lederberg, 1947) will no doubt correct this second shortcoming in due course. But as matters stand at present, the interpretation of 'training' adaptations is controversial, the controversy being between those who maintain that training is secured both by population selection and by heritable transformations of individual cells, and those who maintain that only the former mechanism is at work.

REFERENCES

Abercrombie, M. (1952). *New Biology*, **13**, p. 117.

Agar, W. E., Drummond, F. H., and Tiegs, O. W. (1935). *J. exp. Biol.*, **12**, p. 191.

(1948). *Ibid.*, **25**, p. 103.

Beale, G. H. (1952). *Genetics*, **37**, p. 62.

Beer, G. R. de (1951). Embryos and ancestors. 2nd Ed., Oxford.

Bernheimer, A. W., and Harrison, J. A. (1940). *J. Immunol.*, **39**, p. 73.

(1941). *Ibid.*, **41**, p. 201.

Billingham, R. E., and Medawar, P. B. (1948*a*). *Heredity*, **2**, p. 29.

(1948*b*). *Brit. J. Cancer*, **2**, p. 126.

(1950). *J. Anat., Lond.*, **84**, p. 50.

Brambell, F. W. R., Hemmings, W. A., and Henderson, M. (1951). Antibodies and embryos. London.

Crew, F. A. E. (1936). *J. Genet.*, **33**, p. 61.

Delbrück, M., and Bailey, W. T. (1946). *Cold Spring Harb. Symp. Quant. Biol.*, **11**, p. 33.

Drew, J. S. (1939). *Nature*, **143**, p. 188.

Finlay, G. F. (1924). *J. exp. Biol.*, **1**, p. 201.

Fisher, R. A. (1930). The genetical theory of natural selection. Oxford.

Ford, E. B. (1940). *Ann. Eugen.*, **10**, p. 241.

Guyer, M. F., and Smith, F. A. (1918). *J. exp. Zool.*, **26**, p. 65.

(1920). *Ibid.*, **31**, p. 171.

(1924). *Ibid.*, **38**, p. 349.

Haldane, J. B. S. (1951). *XIX Congr. Int. Philosophie des Sciences, VI, Biologie*, p. 39.

Harrison, J. A., and Fowler, E. H. (1945). *J. Immunol.*, **50**, p. 115.
(1946). *J. exp. Zool.*, **101**, p. 425.
Harrison, J. W. H., and Garnett, F. C. (1926). *Proc. Roy. Soc. B*, **99**, p. 241.
Harrison, J. W. H. (1928). *Proc. Roy. Soc. B*, **102**, p. 338.
Hughes, A. W. McK. (1932). *Proc. Roy. Soc. B*, **110**, p. 378.
Huxley, J. S., and Carr-Saunders, A. M. (1924). *J. exp. Biol.*, **1**, p. 215
Ibsen, H. L., and Bushnell, L. D. (1931). *J. exp. Zool.*, **58**, p. 401.
(1934). *Genetics*, **19**, p. 293.
Jones, F. Wood (1943). Habit and heritage. London.
Kimball, R. F. (1947). *Genetics*, **32**, p. 486.
Kilkenny, B. C., and Hinshelwood, C. (1951). *Proc. Roy. Soc. B*, **139**, p. 73.
Loeb, L. (1945). The biological basis of individuality. Springfield, Mass.
McDougall, W. (1927). *Brit. J. Psychol.*, **17**, p. 267.
(1930). *Ibid.*, **20**, p. 201.
(1938). *Ibid.*, **27**, p. 321.
McDougall, W., and Rhine, J. B. (1934). *Brit. J. Psychol.*, **24**, p. 213.
Medawar, P. B. (1951). *New Biology*, **11**, p. 10.
(1952). *Ibid.*, **13**, p. 116.
Sladden, D. (1934). *Proc. Roy. Soc.*, *B* **114**, p. 441.
(1935). *Ibid.*, **119**, p. 31.
Sladden, D., and Hewer, H. R. (1938). *Proc. Roy. Soc. B*, **126**, p. 30.
Sonneborn, T. M. (1949). *Ann. Rev. Microbiol.*, **3**, p. 55.
(1950). *Heredity*, **4**, p. 11.
Sturtevant, A. H. (1944). *Proc. Nat. Acad. Sci.*, **30**, p. 176.
Tatum, E. L., and Lederberg, J. (1947). *J. Bact.*, **53**, p. 673.
Thomsen, M., and Lemche, H. (1933). *Biol. Zbl.*, **53**, p. 541.
Thorpe, W. H. (1938). *Proc. Roy. Soc. B*, **126**, p. 370.
(1939). *Ibid.*, **127**, p. 424.
Waddington, C. H. (1952). *Nature*, **169**, p. 278.
Woodger, J. H. (1952). Biology and language. Cambridge.

5

The Pattern of Organic Growth and Transformation

'Growth' is a word of notorious imprecision, but it stoutly defies semantical reform. It may mean increase of length, area, weight or volume; it may mean the act or accomplished fact of reproduction, i.e. increase of number; or it may simply mean development—the adverb is not well chosen—with all that development implies of increasing complexity and elaboration. I shall restrict growth here to its simplest meaning, change of size, but I shall consider also the changes of shape which are the outcome of inequalities in the rate of change of size.

Organic growth is not a process of accretion, nor does it build upon an enduring frame. The molecular fabric of the body enjoys no substantive permanence whatsoever, a truth which came to be known in the following way.

The body makes no distinction between the common elements and their various mutants; the natural isotopes of nitrogen and carbon, which have atomic weights of 15 and 13 instead of 14 and 12, or the radioactive isotopes of sodium (24) or carbon (14) which arise by gaining neutrons or losing protons, are exchanged indifferently for their common or parental forms. The administration of compounds containing isotopes distinguished by their mass or radioactivity has therefore made it possible to trace atoms in their passage through the body, and so to reveal the constant exchange of its molecular ingredients for new arrivals from the world outside. Even teeth and bone are the subjects of a restless atomic transubstantiation. It is only the *form* of the body, the system of preferred stations for the inward-bound replacements, that achieves any kind of permanence at all.

Superimposed on these exchanges are the processes which make good the constant wastage of effete or expended cells.

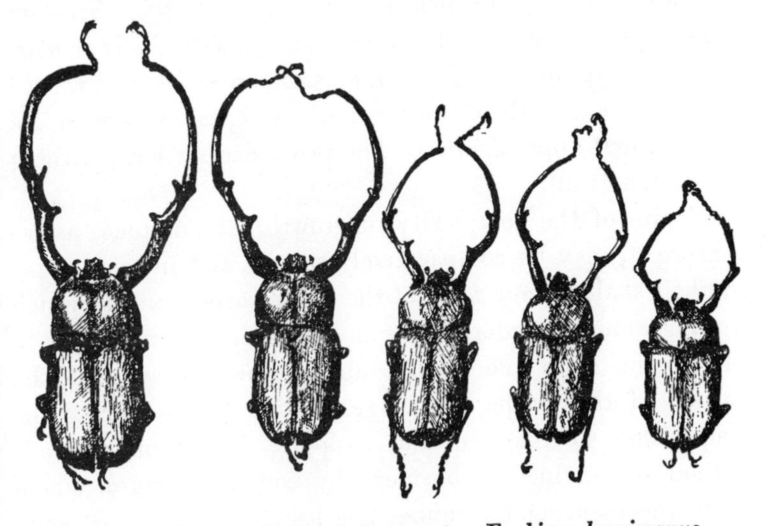

Fig. 6. Male beetles of the species ***Euchirus longimanus***, illustrating how the proportions of an organism may change with its absolute size. The length of the fore-limbs is grossly out of proportion to the length of the body.

Pounds of dead cells in the form of scurf and its several variants (hair, horn, nails, claws) are parted with in a lifetime. The living outer layer of human skin renews itself completely about once a month, or about 100 times in a proverbial seven years. Red blood corpuscles live only about 120 days; at least some lymphocytes appear to be excreted through the walls of the intestine; a small proportion of the finest nerve fibres and blood vessels is probably always in course of disintegration and therefore always in course of being formed anew Replacements of this kind are part of the ordinary maintenance charges of the body: they are not accompanied by any net change of size. But some forms of wastage are integral with the act of growing. Bony tubes and boxes like the long bones of the legs and the cranium are hollowed out on the inside in the course of growing larger. The only growth which is purely additive or accretionary is that of which the product takes no further part in the physiological activity of the body, as with shells or hair. The idea that the growth of organisms can be likened to, for example, the growth of houses is not acceptable even in the roughest first approximation. The two processes have nothing in common at all.

In spite of the complexity of growth, its outcome, as we measure it, may be comparatively simple, and in later paragraphs I shall set out some of the quantitative rules to which growing animals conform. The measurements I shall refer to tell one no more and no less about growth than could be learned of the mechanism of respiration by measuring the composition of inspired and expired air, or of a firm's method of conducting business by contemplating a single figure representing its annual net loss or gain. In all such cases we have to do with measuring the final outcome of covert processes of formidable complexity. The measurements are not very deeply informative, but the information which they contain is indispensable.

THE SCALE OF SIZES

The largest adult mammals are about 50 million times larger than the smallest. A fully grown blue whale weighs about 2×10^8 grams; one of the smallest mammals, the long-tailed shrew *Cryptotis parva parva*, weighs only about four. Even when studied under conditions 'particularly conducive towards repose' this shrew ate its own weight of worms and insects daily, and would have died of starvation if food had been withheld for as little as twelve hours.

The scale of sizes to be found in mammals is not exceptional.

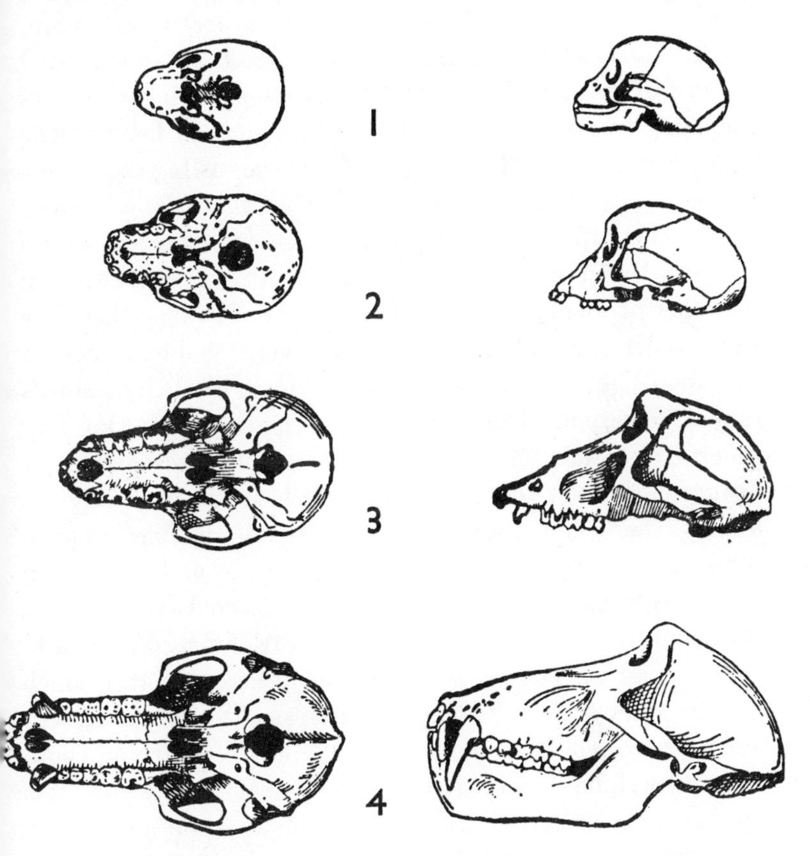

Fig. 7. Differential growth of the skull of the baboon as seen from the under surface and inside view: 1, 2, and 3 are from newborn, juvenile, and adult females, and 4 from an adult male.

The Gobiform fish *Misticthys* is about half an inch long when fully grown and could hardly weigh 1/250th of an ounce. The basking shark (not the largest fish) is known to reach 29 feet in length and to weigh four tons. Dr Harrison Matthews has given excellent reasons for supposing it to be viviparous, though no pregnant specimen has yet been found. The Japanese spider crab may have a claw span of ten feet in extension, but the smallest crustaceans are little more than an animated sea dust in the surface waters of the ocean. The smallest beetles and fairy flies are about 1/100th of an inch in length. The largest squids are 90 feet long and have eyes as big as saucers.

It is not possible to say exactly why animals of a particular species should have come to be of a particular size. The sizes and growth rates of animals are functionally in gear with all the other parameters that define their way of living—their rate and manner of reproduction, their behaviour, habitat, enemies and food. But it is sometimes possible to see why animals cannot be very much larger or smaller than they are. One very general restraint turns on a metrical truism recognized by Spencer—namely, that in a body which is symmetrically enlarging, the volume increases as the cube of the linear dimensions, and the surface area as the square. To multiply length tenfold is to increase surface area a hundredfold and volume a thousand times. In small mammals the ratio of surface area to volume, and therefore the relative rate of loss of heat, is much greater than in large mammals. The smallest mammals eat almost continuously to make good the loss of heat and could not very well be smaller. At the other extreme, the elephant is approaching the upper limit of size for an agile and wholly terrestrial animal. Limbs are roughly speaking strong in proportion to their cross-sectional areas but support a weight that is proportional to volume. The legs of elephants must of necessity be stouter and more pillar-like than the legs of horses; indeed, elephants can be extrapolated for fancy to a size at which one would be lucky to see daylight between their legs. My colleague Mr Maynard Smith estimates that the upper limit of the weight of a flying vertebrate must be about

30-40 lb., because the power needed for flight increases more rapidly than as the cube of the linear dimensions, and therefore out of proportion to the muscles which provide the motive power. Angels, paradoxically, could therefore not be airborne, as Professor Haldane pointed out some quarter of a century ago.

Spencer's Law is more revealing in its actual breach than in its theoretical observance. Surface area keeps pace with the volume it ministers to by folding and subdivision. The walls of the intestine are deeply folded; the walls of the lungs are a multitude of fine sacs. The cross-sectional area of the blood vessels is thought to increase about 800-fold in the passage from the great vessels by the heart to the capillaries of the tissues. The five million red blood cells in a cubic millimetre of human blood offer a surface area 170 times greater than that of a single corpuscle of the same shape and total mass. Spencer's Law is also flouted by physiological adaptation. If the problem of conserving heat is so acute for the smallest adult mammals, how do their newborn babies cope, which are so much smaller still, and hairless? The short answer is that they do not. New-born mice come to no harm by being left for an hour or two in a refrigerator. Their metabolism is such that they are highly resistant to the effects of being chilled.

Another restraint is that which is set by the tempo of diffusion. In the simplest case a diffusing substance penetrates to a distance proportional to the square root of the time during which it has been diffusing. Distance can therefore only be bought at a disproportionate cost of time, a state of affairs which sets definite limits to the permissible shapes of cells. All active cells or one-celled animals which are large are tubular or flattened, except when like yolky eggs they derive energy from stores of food inside.

I hope the foregoing discussion will have distracted attention from the fact that we are very ignorant of the actual and present influences which govern the size-distribution and growth of any wild animal living under natural conditions. Fish are the only wild animals for which we are approaching a predictive ecological theory of growth rate and size frequency. Nor can this be counted a triumph of abstract scholarly enterprise. The pressure of necessity is behind it; it is less because fish are edifying than because they are edible that we know as much as we do.

THE PATTERN OF CHANGE OF SIZE

In spite of the compass and complexity of growth, and the great variety of different processes that contribute to increase of substance, the passage from germ to adult is an orderly and predictable process. What rules of order does it conform to, and upon what reasoning is prediction based?

It is one thing to devise empirical formulae which describe the growth of the members of one particular species; that is simply a matter of measuring the growth of a sufficient number under conditions sufficiently well defined. It is quite another matter to try to frame general laws of growth which the majority of animals are expected to conform to, and biologists have set about the problem in two entirely different ways.

Some have attempted to arrive at Laws of Growth deductively, starting with certain deceptively inoffensive axioms about the course of metabolism and ending with theorems that purport to describe the way in which all animals grow. I believe that this approach must be classified at present as a scholarly indoor pastime; that it may sometimes lead to acceptably accurate answers is only marginal evidence of the truth of the axioms.

The other way to go about it* is to proceed inductively, by

* [The inductive approach is considered in more detail in my article on 'Size, Shape, and Age', in *Essays on Growth and Form*, ed. W. E. Le Gros Clark and P. B. Medawar (Oxford, 1945.)]

recording as many instances of growth as possible and trying to find out the properties they share in common. May we not affirm, for example, that animals increase in size as they grow older, until growth ceases altogether? We may, of course, but only if room is left for reservations. All animals grow smaller if undernourished—a trivial exception—but some animals (like flatworms, nemertines and colonial sea-squirts) 'de-grow' with a deep-seated anatomical retrogression and may even revert to an embryonic level of simplicity. Negative growth of this kind has a special adaptive value. It is not a significant violation of the law of general increase because negative growth is not a reversal of the processes that led to enlargement, as if metabolism has simply been engaged in a reverse gear. Nor is it a significant exception that men and women are shorter in old age than in the physical prime. It is indeed so, and Morant is satisfied that this shrinkage is not just an actuarial artifact due to an earlier death of taller people, nor to the fact that the older people we measure to-day were born longer ago than their juniors and therefore in perhaps less propitious times for growing. (Morant finds that Englishmen a hundred years ago reached the same maximum height as they do at present, but took about five years longer to achieve it.)* Loss of height is probably due to a shrinkage of intervertebral discs. This, too, is not a reversal of synthetic processes; and it may be observed that the luxury of living to an age at which one can indulge in physical deterioration is an artificial by-product of domestication, and a state of affairs that has no parallel in the world of animals at large.

It has long been recognized that biological growth is multiplicative in style, and not accretionary or additive. That which results from biological growth is itself endowed with the power of further growing. In the general case the progeny of a cell

* [Dr J. M. Tanner has since told me that there is clear evidence of a genuine secular increase in height as well as in growth rate.]

(or chromosome or organism) which has divided into two are themselves capable of division, and so in turn their issue. Accretionary products like shells and hair are made by living cells which grow in the organic style: all additive growth is subsidized by acts of multiplication.

A lineage of cells that perpetuated itself without loss by repeated binary divisions would of course increase in numbers in an exponential or geometrical progression. In real life, not even bacteria will increase at such a rate for long. Their growth is restrained by a variety of density-dependent factors, like the accumulation of inhibitory waste products or the exhaustion of the supply of food. Nevertheless, growth by continuous compound interest is the norm for all living systems. It is *departure* from exponential growth that calls for comment and explanation, just as with departure from uniform motion in a straight line. No moving object left to itself will persevere in constant linear motion, and no real organism will grow at a constant specific rate. The former circumstance no more derogates from Newton's First Law of Motion than the latter from what is sometimes called the Law of Malthus. What we must ask is, in what way does the growth of organisms depart from that regimen of continuous compound interest by which they are theoretically empowered?

No one has yet improved upon the answer given by the American anatomist Minot. Consider a sum of money invested at a rate of compound interest which, instead of remaining constant, falls; and let the interest be (say) 10 per cent in the first year, 9 per cent in the second, and 8·1 per cent, 7·3 per cent, 6·6 per cent . . . in successive years thereafter. The sum does indeed grow at compound interest, but the rate of interest falls progressively at a rate which progressively falls. This is the organic style of growth as Minot saw it. A living system progressively loses its power to multiply its substance at the rate at which that substance itself was formed. Put otherwise, the specific acceleration of growth is always negative, but it climbs towards zero as growth proceeds. It is only superficially a paradox that deterioration is faster in young animals than in their elders. Almost all metabolic processes go faster in

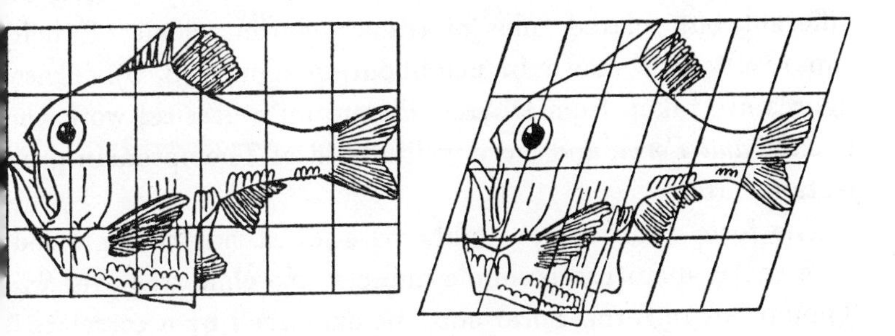

Fig. 8. Line drawings in side view of two related species of small marine fish allotted to different genera, *Argyropelecus* (left) and *Sternoptyx* (right). Plotted on a changed system of co-ordinates, the outline of the one gives an excellent approximation to the outline of the other.

youth than in maturity, and the processes which slow down physiological activity are no exception. We are all moving towards our graves, but none so fast as they who have farthest still to go.

THE PATTERN OF CHANGE OF FORM

Change of size is almost always accompanied by transformation; growth by proportionate enlargement is very rare. We are not born as miniature adults. Inspected through a magnifying glass, a child does not look like a backward adult, but simply like an uncommonly large child.

The shape that is characteristic of adults is governed by spatial inequalities of growth rate, i.e. growth that goes at different rates in different parts of the body. Dr J. S. Huxley has made a special study of these inequalities, and two examples of change of shape cited in *Problems of Relative Growth* are shown in figs. 6 and 7. Adults of different but related species acquire their distinctive shapes because they conform to different but related rules of transformation. Some of their end results are shown in neighbouring figures (8, 9). These figures are taken from D'Arcy Thompson's classical work *On Growth and Form*, and more will be said of Thompsonian projections later.

The form of an object, unlike its size, cannot be expressed by a scalar quantity, a simple number. No child was ever 2·5 Thompsons in form. Form must be expressed by a correlated system of vectorial measurements, i.e. measurements which take account of the disposition of the measured lengths in space. But although shape is in a purely metrical sense indefinable, change of shape is not. Consider a lantern slide thrown on a screen that lies in its normal position at right angles to the optical axis of the projector. When the screen tilts one way or another, the cone of light is cut at different angles and the image is accordingly transformed. The nature and degree of the distortion can be expressed with mathematical exactness. No matter how complex the pattern of the image, its change of shape can be accurately defined.

Huxley's method of assaying change of shape in development is to measure the growth rate of one part of the body in terms of the growth rate of another. If transformation is an orderly process, the two sets of measurements will vary in dependence on each other. In the simplest case, not uncommon but by no means universal, the parts in comparison multiply their sizes in a constant ratio: the size of one is a fixed multiple of the size of the other when the size of the other is raised to a constant power. Proportions alter, therefore, but alter in geometrical progression. It is only when the ratio or power is unity that the proportions of the growing parts stay constant, and this, as I have said, is rare.

It follows that just as a growing animal must traverse all

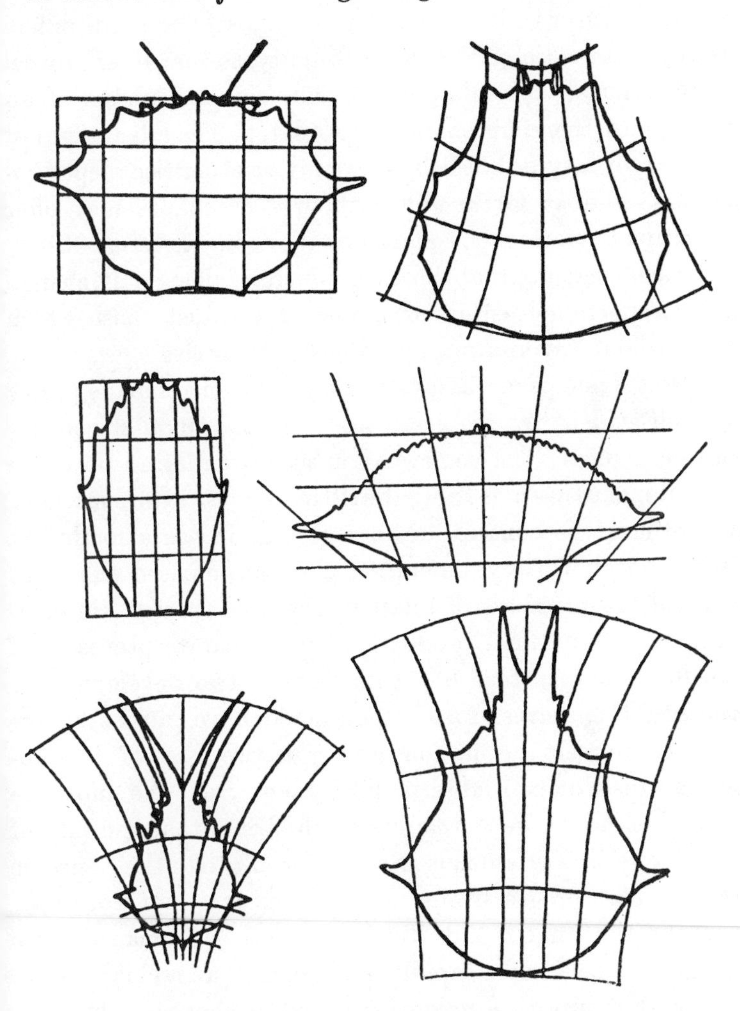

Fig. 9. The carapaces of crabs of six different but related genera, showing how particular differences of form may be expressed as the outcome of a general process of orderly spatial transformation.

intermediate sizes before it attains to adult weight or stature so it must traverse a spectrum of intermediate conformation before it can reach its adult shape. A particular shape can only be 'fixed' if growth itself comes to a standstill or if the differential growth-ratio settles down to unity, so that later growth entails symmetrical enlargement. The size and shape of an animal must therefore be nicely correlated. The advantages of being larger may be offset by unwieldy or otherwise inept proportions, and as far as different sizes or shapes may offer competing inducements, so far must they come to terms. It may not be a coincidence that those fish which, of all animals, change their proportion least in development are just those which grow without any known upper limit to their size.

D'Arcy Thompson's assay of transformations is made pretty well self-evident by the grids superimposed upon the neighbouring figures. The somewhat arbitrary tailoring of space which it makes use of is fraught with metaphysical implications but we must be content to observe that it has a forthright visual appeal. D'Arcy Thompson always compared the *adult* forms of the members of related genera or species. He compared, then, the final products of two separate processes of transformation, instead of comparing the two developmental processes themselves. Ideally he should have put both processes into cinematic motion, giving us two films of development instead of two lantern slides taken from the ends. He could thus have given precision to the belief that the rate of change of shape of animals in development, like their specific growth rate, progressively slows down. It is, of course, a generalization that is 'intuitively' obvious—a human embryo changes its shape more rapidly in a month than a child does in a year—but intuitive judgements are inoffensive only when everyone agrees with them, and in palaeontology, where the problem of the assay of form is ever with us, this is by no means so. But D'Arcy Thompson's method as it stands leads to the important inference that change of shape is orderly not only in time but in its spatial distribution, and that a multitude of particular differences of shape between two living organisms may be only the topical expressions of a single, simple, comprehensive change of form.

The rules of organic transformation are therefore analogous to those we have already arrived at in respect of growth. First, both size and shape change in course of development, and change continuously within the compass of their upper and lower limits. Change of size has a definite sense or trend, viz. of increase, and change of shape has also a definite trend. (Animals do indeed 'increase' in form as they develop, if by that we mean that they increase in order of complexity; but change of complexity is outside the competence of D'Arcy Thompson's method, which must confine itself to homeomorphic forms.) Both then are progressive processes: it is exceptional for animals to grow smaller as they become older, and equally exceptional for them to reverse the prevailing trend of change of shape. Animals pass once through intermediate sizes before they reach adult weight or stature, and once through intermediate shapes before assuming their adult form. The specific growth rate is greater in early life than latterly, and so also is the rate of change of form. It all amounts to saying that growth is orderly in space as well as in its temporal unfolding, and that the ordinances are rather simpler than one might at first suppose.

Acknowledgment is made to the Zoological Society of London for figure 7 (from a paper by Professor S. Zuckerman in *Proceedings of the Zoological Society*) and to the Cambridge University Press for Figures 8 and 9 (from D'Arcy Wentworth Thompson, *On Growth and Form*). Figure 6 is from C. Champy, *Sexualité et Hormones*.

6

The Imperfections of Man

Evolution is one of the subjects upon which laymen have long felt themselves entitled to express an opinion; formerly, the opinion that evolution does not occur; latterly, that it does not occur in the way that biologists now suppose. The chief among several causes of their present discontent is approximately as follows. Biologists believe that evolution has come about through the action of material forces, in the sense that it does not unfold itself according to a preordained purpose or superimposed design. To laymen, an argument which takes no account of design or purpose or Aristotelian Final Causes is utterly unsatisfying and implausible. How can mere unguided material forces be responsible for the miraculous optical engineering of the eye; for the exquisite functional aptitude of a bird's wings; for the almost finicky precision of mimicry? Is it not going a little too far to impute these splendid accomplishments to what Bacon called the 'casual felicity of particular events?'

These are intelligible complaints, but they are founded upon a misconception, namely, that evolution is a perfectionist process. The eye, for example, is beset by chromatic and spherical aberration, and is not correctly centred along its optical axis; Helmholtz, the grand master of physiological optics, said that an optician would be ashamed to make an instrument with such elementary physical faults.* Many of the

* [*Vorträge und Reden*, I, p. 286, 1903. Helmholtz had obviously been exasperated by contemporary 'nature-philosophers' of the perfectionist school: see his *Treatise on Physiological Optics*, American edition, I, p. 185.]

perplexities of laymen might be set at rest if it could be shown that evolution is very much a fallible, makeshift affair, and that loss of fitness in one regard is often the charge for some more-than-compensating gain. I choose the imperfections of man as the subject of this essay, because man's superlative biological status is hardly to be questioned, and shall take three examples of his falls from grace: his susceptibility to haemolytic disease of the newborn; the mechanical shortcomings of his upright carriage; and the ineptitude of wound healing in injuries of his skin. Of these, the first is familiar enough, and I shall only deal with its broader aspects; the second is well understood but not yet widely known; and the third will be unfamiliar to all except a few specialists in the theory of wound healing.

* * * * * *

Haemolytic disease of the newborn is the general name given to a variety of affections (kernicterus, icterus gravis, hydrops fetalis) marked by grave and sometimes fatal abnormalities of the blood and blood-forming organs; its interpretation, which we particularly associate with the names of Levine, Landsteiner and Wiener, is one of the great triumphs of modern clinical biology. Briefly, it is an *immunological* disease; it depends upon the active immunization of the mother against blood group substances (chiefly those of the Rhesus and Kell systems) absent from her own tissues, but present in the tissues of her child. Like many forms of allergy and hypersensitivity, and like the reaction that forbids the use of one person's skin to repair another's, haemolytic disease can be described as a miscarriage of immunological justice—a harmful and apparently wanton aberration of what is properly and primarily a mechanism of defence.

Haemolytic disease of the newborn is a peculiar menace to human beings for the following reasons. If it is to occur at all, two qualifying conditions must be satisfied at the outset. First, the antibodies which are the chemical effectors of the immunity reaction must be able to pass from the mother's circulation into the circulation of the unborn child. In effect, this means that the membranes which separate mother from foetus must be of such a kind as to let the antibodies through. This first condition is satisfied by rabbits, rats and mice, and also by human beings, but not by cattle, sheep and horses. Second, the foetus must reach before birth a stage of development at which the immunizing substances present in its blood corpuscles have reached maturity, for if they are still undeveloped at birth the mother can have no normal opportunity to become immunized against them; and even if the mother were to be artificially immunized, the antibodies so formed could not attack the foetal blood. This second condition is satisfied by cattle, sheep and horses, and also by human beings, but probably not by mice and not by rats.

On the face of it, mice should be specially liable to be immunized by their own young, for a female mouse may give birth to a quarter of her own body-weight of young in a single pregnancy and to ten times her body-weight in a lifetime; furthermore, antibodies formed in the mother have ready access to the embryos within it. But it seems to be impossible to give mice haemolytic disease, even when their mothers are deliberately immunized against the red blood corpuscles of their young. Mitchison attributes this to their extreme immaturity at birth; if the degree of maturity of red blood cells is taken as a yardstick, birth in mice takes place at a stage equivalent to a human embryo that is still six months from term. Conversely, cattle, horses and sheep can only get haemolytic disease after birth—a stage at which it is clinically manageable—for it is only at birth that antibodies pass from the mother to her young, in the colostrum, the first watery milk.

Human beings qualify on both counts: maturity at birth and the ability of antibodies to gain access to the unborn young.

Nor is this the whole story. Haemolytic disease can only occur if the members of an interbreeding population are dissimilar in their antigenic make-up. If all human beings were Rhesus-positive or Rhesus-negative, it is obvious that they could not get haemolytic disease (or suffer from transfusion accidents) as a result of immunization by the Rh antigens. But they are, as it happens, most highly diverse with regard to the antigens present in their red blood corpuscles. About one in six Englishmen lack the most mischief-making antigen of the Rh series, viz. D or Rh_0; about one in ten possesses the Kell antigen.

We may now ask, what advantage do human beings enjoy which compensates, or more than compensates, for their vulnerability to haemolytic disease? It is clear that man's embryological advantages, if such they are—a long gestation period, coupled with a form of gestation which allows beneficial as well as harmful antibodies to enter the foetus—could be enjoyed with impunity if the entire population were either all positive or all negative in respect of the blood group substances Rh and Kell. The question therefore resolves itself into asking: Why, then, have they not become so? Unfortunately, the answer is not known; it is merely being groped after.

Roughly speaking, there are two kinds of reasons why a population should be polymorphic, i.e. should be subdivided into variant types of which even the least frequent is far too frequent to have originated merely through the recurrence of mutations. The first is that heterozygotes (which carry and therefore propagate a gene, but do not necessarily reveal its presence) should stand at some special advantage relative to homozygotes. Contrary to all superficial appearances, this appears to be true of the blood affection known as 'sickle-cell trait', the analysis of which provides a most noteworthy example of a combined operation in genetics, chemistry, anthropology and clinical medicine.

Briefly, 'sickle-cell' trait is an affection in which the red cells of deoxygenated blood adopt a sickle shape. Pauling and his colleagues showed that it is due to the presence of an abnormal variant of adult (as opposed to foetal or infantile) haemoglobin. Sickle-cell trait appears in about 9 per cent of American negroes, and in a proportion varying from 0 to 45 per cent in communities of West African negroes. Its inheritance, worked out independently by Neel and Beet, is governed by a Mendelian dominant gene, and sickle-cell trait (a condition which is not harmful in itself, nor even appreciably disabling) is its heterozygous or hybrid form of expression. Individuals who are homozygous for the sickle gene, however, suffer from a grave and sometimes fatal anaemia, and few of its sufferers live to reproduce.

In the face of powerful selection against the homozygous form the frequency of the sickling gene is far greater than can be accounted for by mutation. According to Allison, the reason why it flourishes is that the heterozygote is endowed with a specially high resistance to subtertian malaria; in areas of the world where malaria is hyperendemic the gene is therefore kept in being by the high selective advantage of the heterozygote; elsewhere it is going or gone.

No such neat and rounded story can be told of blood group polymorphism; indeed, it was at one time thought (and by geneticists feared) that the subdivision of human beings into the blood groups A, B, AB, O was entirely capricious, in the sense that an individual's blood group had no bearing on his fitness to survive and reproduce. Workers at the British Post-graduate Medical School now find that membership of blood group A is associated with an increased susceptibility to cancer of the stomach, and membership of group O with a greatly increased susceptibility to peptic ulceration. Blood group polymorphism is thus certainly not a matter of indifference, though its import is still obscure. But, so far as I am aware, no one has yet been able to associate the subdivision into Rh blood types with anything except the unqualified incubus of transfusion accidents and haemolytic disease.

It is possible, then, as Haldane has suggested, that the diversity of Rh blood types represents a second kind of polymorphism—that which is merely transient, a necessary intermediate stage between the elimination of one type or the other. This interpretation is borne out by the existence of rather bold inequalities among different races and nations in the proportions which belong to the several groups. The number of Basques who are Rh-negative falls between one in three and one in four, but Levine and Wong found only one Rh-negative individual among 150 Chinese, having been led to their enquiry by the significant observation that haemolytic disease in China is very rare. If the interpretation is true, then haemolytic disease could be explained away as a transient genetic ailment of mankind, but fortunately we can look forward to something a little more expeditious than an evolutionary cure.

* * * * * *

Man's upright carriage may be a constant source of moral satisfaction, but it has certain serious mechanical drawbacks. Man is unique among four-legged animals in being able to stand erect, on the flat of his feet, and to balance himself in that position. (Even gorillas do not stand upright more than momentarily, and they walk not on the flats but upon the outer margins of their feet.) The shape of the backbone has changed accordingly. In all other animals, with unimportant exceptions, the backbone is more nearly horizontal than vertical, and it takes the form of a single unkinked or uninflected arc from neck to tail. The 'vertebral column' is not a column at all, but is more like a cantilever having the four legs as piers. The vertebral column of a human being is no longer a simple uninflected arc; it bends slightly forwards in the neck, slightly backwards in the thoracic cage; forwards again in the lumbar region, the small of the back, and backwards in the fused vertebrae that form the sacrum. That is the mature pattern; in development, the neck flexure appears somewhat before birth, and the lumbar flexure between the ninth and eighteenth months of age.

An upright stance imposes new and peculiar stresses upon the spinal column. The support of weight imposes a force acting down the vertical long axis, which tends to compress the vertebrae upon themselves. The angle of their apposition is responsible for a shearing force between the bottom-most lumbar vertebra and the sacrum; and general flexional strains become very apparent when stooping to pick up weights. To cope with these new forces (for such, in an evolutionary sense, they are) man inherits only the standard outfit of muscles and ligaments, and the muscular bracing of the neck and lumbar region leaves much to be desired.

What suffers from the wear and tear of habitual use is not, primarily, the vertebrae themselves, but the tissues lying between them. The bodies of the vertebrae are not set against each other face to face; on the contrary, about one-quarter of the total height of the column (more in the lumbar region) is occupied by peculiar solid intervertebral joints. Each joint forms a so-called intervertebral disk—a central nucleus of semi-fluid consistency, which embodies or represents the remnant of the embryonic notochord; contained within a tough fibrous ring, the annulus, in which the fibres are disposed cylindrically in coaxial rings; the whole being bounded above and below by flat cartilaginous plates. The whole organ has been described by one of its leading students, Ormond Beadle, as a hydro-dynamic ball bearing.

The bearings may give in a variety of different ways. Under repeated flexion of the spinal column when it is taking weight, the nucleus may gape through a weakness in the fibrous ring which normally contains it, press against the posterior ligament, and even encroach upon the spinal canal. Alternatively, as if by the insistent action of 'telescoping' forces, nuclear matter may break through perforations in the cartilaginous plates and obtrude into the vertebral bodies, which are made of spongy rather than of compact and concentrated bone. In recent years, the opinion has been gaining ground (perhaps too rapidly) that many disabilities which have been loosely classified as sciatica, lumbago and vague rheumaticky back pains are due to abnormalities of the intervertebral disks; the immediate causes of pain may still be debatable—for example, as to whether or not mere chronic pressure on a nerve root can cause inflammation and thereby pain—but their anatomical origin seems pretty certain.

The exhaustive anatomical studies of the Dresden pathologist Schmorl have led the way to a conclusion of more general biological interest, that the spinal column is the first organ in man to 'age', that is, to show the deterioration consequent upon ageing. Pathological changes, it has been said, are detectable as early as the eighteenth year of life. The deterioration of the spinal column provides, indeed, what is perhaps the best example of a process of ageing which at least *begins* by being a consequence of the cumulative effects of wear and tear —of chronic functional attrition, as opposed to the 'innate deterioration', which takes place independently of abuse, or even use.

Man's upright stance has incomparable advantages, perhaps above all in providing for his principal physical (as opposed to mental) asset, manual dexterity, but it takes its toll in the mechanical vulnerability of the spine. 'Disk lesions' are not, of course, peculiar to man, but it does not derogate from our argument that they should also occur in dogs, for they are found principally in those breeds (bull-dogs, pekinese) which have been deliberately selected for imperfect cartilaginous development. Only we ourselves, therefore, not natural evolution, can be held to blame.

* * * * * *

The third of man's peculiar shortcomings on our agenda is the appalling ineptitude of wound healing in the skin, and here I shall follow closely the reasoning of Billingham and myself. The following comparison will make the problem clear.

If the entire thickness of the integument in the chest region of an adult rabbit is excised over a rectangular area of 100 cm.^{2}; something that looks superficially like an irreparable injury is produced. But, so far from being irreparable, it requires for its quick and successful healing nothing more than the most elementary surgical care. The surface area of an adult human being is about seven or eight times as great as a rabbit's, but a skin defect of the same absolute size and depth, and the same relative position, cannot by any means be relied upon to heal satisfactorily of its own accord. If left to itself, it will heal painfully slowly, and will gather up and scar; a wound of similar size in the leg (which is not so much thinner than a rabbit's trunk) could cause a seriously disabling injury if left untreated, whether by gathering up in such a way as to constrict the blood supply of the limb or by immobilizing a joint. Such an injury cries aloud for skin grafting, an operation in which a thin flat slice of normal skin is removed from some undamaged part of the patient's body and held in place for four or five days over the area of loss. (The skin graft is removed so thinly as to leave behind part of the leathery layer of the skin of the donor area, and the bases of the hair shafts; the donor area will therefore heal of its own accord without scarring or contraction.)

The question is, why is the rabbit so accomplished in wound healing and the human being so strikingly inept; the answer turns upon an understanding of the mechanism of healing as it occurs in a rabbit's skin.

As Billingham and I see it, there are two quite distinct (though concurrent) elements in the healing of rabbits' skin. The first is contracture, which closes the wound by a progressive coming together of its original edges. Contracture follows a regular geometric pattern: starting as a rectangle, the wound first becomes smaller without changing shape; then the sides cave in towards the centre, and meet from the four corners inwards, so that all that is left is a neat >—< shaped line of suture (it does not deserve to be called a scar). During the process of contraction the raw wound area is temporarily covered by a thin film of skin epithelial cells which grow or migrate inwards from the edges of the wound; but as the original skin edges come together, so the space enclosed by them diminishes and finally disappears, and the skin epithelium, which is purely a temporary organ of healing, disappears with it.

Contracture closes the wound, but it does not, of course, make good the loss of 100 cm.^{2} of skin. Billingham and I believe that this loss is made good by a second process, the intercalary or 'intussusceptive' growth of the remaining skin. Intercalary growth is an expansion of the skin by growth on or within its existing fibrous framework. The simplest way of demonstrating it is to prepare a rectangular wound and to leave behind, in its centre, a small area of undamaged skin. Alternatively, skin may be excised from the whole of the area and a skin graft thereupon placed in the middle. The forces of contracture which draw the skin edges together bring an expansive force to bear on the central island, which may accordingly enlarge to no less than ten times its original area. The number of hair roots is not added to,[1] so that the most obvious outward evidence of intercalary growth is the fact that the hairs growing from the central island become spaced apart to a degree proportional to its linear enlargement. There is nothing particularly novel or mysterious about the intercalary enlargement of skin, for it is a process that occurs naturally in growth from newborn to adult size. It is made particularly obvious by the fact that when a child is grafted with skin, the graft grows with him.

Contracture and intercalary expansion between them make for admirably efficient healing in rabbits. Why do they not work to equally good effect in men?

The answer is probably anatomical. Contracture can only be an efficient healing process if the skin is sufficiently loose to 'give' while it is going on. In rabbits, and in mammals generally, the integument is very loosely knit to the body wall; its main blood supply runs in a plane parallel to the surface, and it contains its own intrinsic musculature (the 'panniculus carnosus') which makes it possible for mammals to twitch their skins. In human beings, the integument is no longer a generously fitting coat, but is much more firmly knit to the tissues below; the intrinsic muscles of the skin are now confined to areas of the face and neck, and the skin generally is much more of a piece with the rest of the body. The upshot of this new anatomical arrangement is that contracture, so far from being an efficient mechanism of wound closure, has become something of a menace; it constricts, disfigures and distorts, and may yet fail to bring the edges of the wound together. *But it still occurs*:

[1] Dr Breedis has recently shown that, contrary to almost all expert belief, the number of hair roots *can* be added to in adult animals, and my colleagues R. E. Billingham and Paul S. Russell have done experiments which entirely bear out his views. The new formation of hair roots, however, takes place only under certain special circumstances that do not affect experiments of the kind described above.

human beings may be said to have retained the mechanism of healing by contracture, but to have lost the anatomical prerequisites which enable it to proceed to good effect. As a mechanism of wound healing, the contracture of human skin is therefore as archaic as the vermiform appendix; like the appendix, we become aware of it only when it leads to harm. Fortunately, thanks to the ingenious and entirely artificial act of skin grafting, human beings need no longer suffer the dilatory and incompetent ministrations of the 'natural' process of repair. What compensating advantage the human being gets from the novel structure of his skin is far from obvious, though it is hard to believe that there is none.

* * * * * *

What I have sought to show in this article is that evolutionary advancement is a compromise between what is desirable in the abstract and what can in fact be done; that the lesser evil must be put up with if it makes possible the greater good; and that bad mistakes are made which, though foreseeable to a prescient mind, were not in fact foreseen. The philosophic import of this proposition may well be most debatable, but the truth of the proposition itself can hardly be in doubt.

7

Tradition: The Evidence of Biology[1]

In order to avoid any possible misunderstanding, I want to make it clear from the beginning that I am going to address you as a professional biologist, and that I shall consider only that fraction of human behaviour about which a biologist might be expected to have something pertinent to say. I shall touch very briefly upon two problems. First, what is to be learned about the causes and motives of human behaviour—about our Springs of Action—by thinking of man as 'just another animal', that is, by thinking of the biological similarities between animals and men? Everybody recognizes that there are indeed profound similarities between the behaviour of man and animals, but biologists and laymen think about them in entirely different ways. When laymen see mice nursing and cherishing their young, their first thought is 'How like human beings they are, after all!' The biologist (at all events when he is on duty) thinks 'How mouse-like, after all, are men!'

The second question is, what is to be learned by reflection upon the biological *differences* between men and other animals? In answering this question I shall come to a conclusion that may surprise you, viz. that tradition is responsible for a large part of the present biological fitness of man.

[1] [The gist of one of several short addresses on 'Tradition' given at a Present Question Conference in 1953 on the general theme of *Springs of Action*.]

I

In everyday life (as opposed to conferences) we never think about Springs of Action in a general way at all; we think only about the springs of *particular* actions and with the problem of choosing between one action and another. We do not worry about why human beings have propensities for loving and hating, but about why one person loves a second and hates a third. We take it for granted that people need food and take steps to get it, but what is *interesting* about food-seeking activities of human beings is why they eat this and not that, here and not there, now and not then. It is no great new truth that human beings are ambitious; what is interesting about ambition is why in one person it takes the form of wanting to become a great musician, in another of wanting to raise a large family, and in a third (for this too is an ambition) of wanting to do nothing at all. In these three examples I hope you can see a clear distinction between the propensities underlying certain general kinds of behaviour and the factors which decide that a certain general kind of behaviour shall take a certain particular form.

Unfortunately, the evidence of biology does not yet run to analysing the sources of particular human actions and decisions: that is a matter for psychology, or perhaps for common sense. But that does not mean that the evidence of biology is uninformative or dull. Tinbergen[1] and Lorenz have given us reasons for believing that many kinds of behaviour which seem to us to be peculiarly human are part of a very ancient heritage —'showing off', for example; playing with dolls; sexual rivalry; and many kinds of 'displacement activity', in which a thwarted instinctive impulse vents itself in actions of an apparently quite

[1] See N. Tinbergen's *The Study of Instinct*, Oxford, 1951. My indebtedness to Tinbergen will be very obvious to anyone who follows the newer research on animal behaviour ('ethology', as it has come to be called).

irrelevant kind. In the main, though, the evidence of biology serves only to identify the parameters in the equations of human behaviour, if 'instincts' can be so described.

It is not at all easy to define instinctive behaviour, but it has certain properties that distinguish it clearly from behaviour of other kinds. First, instinctive behaviour is *unlearned.* In practice, it is sometimes very difficult to decide whether a particular act of behaviour is learned or unlearned, and with human beings it may be well-nigh impossible. With animals it is simple enough in principle. Is nest-building activity in mice and birds copied or inborn? To answer this question decisively, one must rear a mouse or bird in strict isolation from birth, away from any possible source of information about how nests are made; and one finds that a mouse so reared can make itself an admirable nest. This provides one good reason for describing nest-building as an instinctive act. Then again, instinctive activities are 'purposeful'. Never mind the teleological undertones of the idea of purpose; all I mean is that instinctive activities, however complex, do in fact converge upon a certain goal—mating, feeding, or whatever the case may be. The functional import of instinctive activity, and the pattern of the connections between its several parts, is intelligible only by reference to the goal towards which it is directed. Thirdly, instinctive behaviour has a 'drive', a sort of psychological pressure behind it; a drive, for example, to find food or to find a mate. The drive is temporarily discharged or assuaged by the act which constitutes the goal of a particular instinctive action. It is usually possible to distinguish two phases in a sequence of instinctive behaviour: first, 'appetitive behaviour', that earlier phase in which an animal seeks the means of gratifying its instincts; and second, the performance of the 'consummatory act' which finally achieves the goal. The appetitive behaviour that gives evidence of a hunger drive includes all the activities entailed by seeking food. The consummatory act of the instinct is eating, and with its performance the hunger drive is discharged or worked off, and that particular episode of instinctive activity comes to an end.

Students of animal behaviour have described, analysed and then pieced together again a great variety of different kinds of instinctive action. Two conclusions which can be drawn from their work, though both are negative, have a profound bearing on human affairs. There is no such thing as an 'aggressive instinct', and it is therefore altogether wrong to suppose that human beings can be its victims or its beneficiaries. There is no drive, no motive force in animal behaviour that is discharged or gratified by the mere act of fighting. Fighting and aggression—much of it bluff—do indeed play a part in animal life, but they are entirely subsidiary or incidental to certain other complex instincts. Males may fight in establishing their mating territories, and fighting may play a part in seeking food or in defence, but there is no such thing as an 'aggressive instinct' in itself. There is simply an aggressive element in several instincts, as there may well be an element of co-operation or mutual aid. There is equally no such thing as a 'social instinct', no sort of inward compulsion that is set at rest merely by getting together in groups; but co-operation is, as fighting is, a component of several different kinds of instinctive behaviour.

Let me say just one other thing about the rôle of instinctive activities in human conduct, using this quotation from Alfred North Whitehead as a text:

> It is a profoundly erroneous truism, repeated by all copy books, and by eminent people when they are making speeches, that we should cultivate the habit of thinking what we are doing. The precise opposite is the case. Civilization advances by extending the number of important operations which we can perform without thinking about them.

This is a most important and arresting half-truth—so compellingly true, or half true, that one wonders how anyone could ever have held a contrary opinion. But it is true only of *learned* activity. No matter what the activity may be—learning the multiplication table, or how to drive a car, to speak intelligibly, or to sew—learning is a process of thinking and deliberation and trial and decision, but the *state of having learned* is the state in which one need think no longer. Paradoxically enough, learning is learning not to think about operations that once needed to be thought about; we do in a sense strive to make learning 'instinctive', i.e. to give learned behaviour the readiness and aptness and accomplishment which are characteristic of instinctive behaviour. But that is only half the story. The other half of the half truth is that civilization also advances by a process which is the very converse of that which Whitehead described: by learning to think about, adjust, subdue and redirect activities which are thoughtless to begin with because they are instinctive. Civilization also advances by bringing instinctive activities within the domain of rational thought, by making them reasonable, proper and co-operative. Learning, therefore, is a twofold process: we learn to make the processes of deliberate thought 'instinctive' and automatic, and we learn to make automatic and instinctive processes the subject of discriminating thought.

II

I now want to try to answer the second question I put before you: in what fundamental biological way do human beings differ from other animals? One possible answer, which I shall try to justify, is this: man is unique among animals because of the tremendous weight that tradition has come to have in providing for the continuity, from generation to generation, of the properties to which he owes his biological fitness.

It is the merest truism that man is a tool- or instrument-using animal. The instruments used by human beings are of two chief kinds. The first I shall call motor or effector instruments—for example, hammers, cutlery, motor-cars, megaphones and guns, instruments which increase, sometimes prodigiously, our repertoire of motor activity. Instruments of the second kind can be described as sensory accessories: spectacles, ear-trumpets, radio sets, thermometers, appliances which increase beyond all former bounds the competence of one's ordinary senses. (Not all instruments fall into these two categories: clothes, for example, do not.) Man is not quite uniquely an instrument-using animal; but the odd examples of the use of tools by lower animals are so rare that each one is treasured and made a fuss of. The Galapagos woodpecker, a sort of finch, uses a thorn held in its beak to prise insects from the bark of trees. Many animals make houses or shelters, but these are tools of a kind I shall not be concerned with here.

I propose to use the terms invented by the great actuary A. J. Lotka to distinguish between the organs that we are born with and organs that are made: *endosomatic* instruments for eyes, claws, wings, teeth and kidneys, *exosomatic* instruments for telescopes, toothpicks, scalpels, balances and clothes. Although there is a very obvious distinction between instruments of these two kinds, the distinction is much less obvious biologically than it is to the unaided power of common sense. The two kinds of instrument serve the same biological functions, and each can to some extent deputize for the other. Even in a quite narrowly biological sense, man is a flying animal: he can fly faster, farther and higher than birds, if not yet with quite the same finesse. It is also important to remember that exosomatic instruments are *functionally* parts of the body, even if they are anatomically separate and distinct. All sensory tools like spectacles, Geiger counters and spectrophotometers report back at some stage and by some route through the ordinary endosomatic senses, and motor instruments are functional, and functionally intelligible, only when they are used. It is not spectacles, but spectacles worn and looked through, that are instruments of vision, and the hammer is only a tool when it is wielded by the hand. (I think it was Wilfred Trotter who said that when a surgeon uses a simple instrument like a probe or seeker, which is merely an extension of the fingers as stilts are extensions of legs, he actually refers the sense of touch to its tip.) The relationship between instrument and user may be very remote, as it is with guided missiles and with engines designed to work without attention, but their conduct is built into them by human design and, in principle, their functional integration with the user is just the same. It is for this reason I deplore the habit of describing the brain as a kind of calculating machine; the truth is that a calculating machine is a kind of exosomatic brain. It performs brain-like functions, much as cameras have eye-like and clothes have skin-like functions, and motor-cars the functions endosomatically performed by legs. We may indeed learn something about the brain by studying calculating machines, as we have learned something about the eye by studying lenses; but it need not be so: the internal-combustion engine has no lessons to teach us about how muscles work.

Biologists in the nineteenth century were much impressed by the fact that exosomatic instruments undergo a systematic secular change that is somewhat analogous to ordinary biological evolution. Just as legs and ears have changed in the course of time, so also have bicycles, microscopes, radio sets and cars. The evolution of both endosomatic and exosomatic organs is gradual in synoptic view, but somewhat discontinuous on closer inspection. Novelties arise in both, not in the entire population but in a limited number of its members, and may or may not spread thereafter through the whole. Both modes of evolution are 'integrative' in the sense that they start or proceed from the groundwork of what has been achieved before. Failure, abortion or extinction occur as commonly in the one as in the other; 'vestigial organs', like the coccyx or appendix, or those functionless buttons on the cuffs of men's coats, occur in both. Whether or not you choose to describe the systematic secular change of exosomatic instruments as an 'evolution' is utterly unimportant; all I am concerned to emphasize is that both exosomatic and endosomatic 'evolution' are equally modes of the activity of living things, and that both are agencies—to some extent alternative agencies—for increasing biological fitness, i.e. for increasing those endowments which enable organisms to sustain themselves in and prevail over their environments. In man, ordinary evolution as we understand it in lower animals, endosomatic evolution, does still happen, and I could give examples of evolutionary changes that have occurred within the known history of the human race. But they are changes of a comparatively minor character, whereas the changes wrought upon human society by exosomatic evolution have been rapid and profound.

I now at last come to the point. There is one crucial distinction between endosomatic and exosomatic evolution. Ordinary evolution is mediated by the process of heredity. Exosomatic 'evolution' (we can still call it 'systematic secular change') is mediated not by heredity but by *tradition*, by which I mean the transfer of information through non-genetic channels from one generation to the next. So here is a fundamental distinction between the Springs of Action in mice and men. Mice have no traditions—or at most very few, and of a kind that would not interest you. Mice can be propagated from generation to generation, with no loss or alteration of their mouse-like ways, by individuals which have been isolated from their parental generation from the moment of their birth. But the entire structure of human society as we know it would be destroyed in a single generation if anything of the kind were to be done with man. Tradition is, in the narrowest technical sense, a biological instrument by means of which human beings conserve, propagate and enlarge upon those properties to which they owe their present biological fitness and their hope of becoming fitter still.

8

The Uniqueness of the Individual

1. INTRODUCTION

Philosophy and common sense, though often parted, have long agreed about the uniqueness of individual man. Different men have different faces, sizes, shapes and origins; different aptitudes, skills and predilections; and different ambitions, hopes and fears. Science now makes it a trio of concordant voices, for the uniqueness of individual mice and men is a proposition which science can demonstrate with equal force, perhaps with deeper cogency, and certainly with a hundred times as much precision. For reasons that will not become apparent until later, I shall begin what I have to say with some observations on the repair of burns.

Deep and extensive burns are injuries which cannot heal properly of their own accord. What happens when a severe burn is left to heal by its own devices is, in outline, this. The skin which has been killed by burning eventually sloughs away; its place is taken by a spongy, moist, highly vascular tissue of repair called granulation tissue. Repair, such as it is, is brought about by two concurrent processes. The epithelial cells which form the outermost layer of the skin creep inwards over the surface of the granulation tissue, and as they do so, the granulation tissue begins to form connective tissue fibres similar in individual make-up to those which form the leathery layer of the skin but arranged in a different and functionally ineffective pattern. In the meantime, the edges of the wound are forcibly drawn together by the tensile forces generated within the granulation tissue: this is the process of contracture. In rabbits and other mammals with a loose integument, contracture is the normal mechanism of repair, and it works admirably, for a rabbit's skin is so mobile that the edges of the wound can be drawn together into the neatest possible natural line of suture. Human beings retain the mechanism of healing by contracture, but human skin is so firmly bound to the tissues underneath it that contracture is no longer an efficient method of repair; the edges of the wound are forcibly dragged inwards, instead of giving easily, and the skin around is gathered up and distorted. (Mere disfigurement is one of the lesser evils, for in certain parts of the body contracture can constrict the blood-vessels, acting like a tourniquet, or immobilize a joint.) The end-result of this entirely inept process of repair can only be described as functionally and cosmetically abominable. A greater or lesser scar is left, made of dense fibrous tissue, and covered by a sometimes unstable and always unsightly surface layer of epithelium which never regains its natural suppleness and colour nor grows anew its normal endowment of glands and hairs. But before the surgical innovations I am about to describe came into common use, to achieve even such an end-result as this would be a matter for congratulation, for 'natural' repair is a dilatory process that gives the patient every reasonable opportunity to die from the steady seeping away of body fluids through the wound's raw surface, or from the wound infections that, without the help of antibiotics, an already enfeebled body could do little to oppose.

For this appalling problem, the surgical operation of skin grafting provides an almost complete solution. What is done, as a rule, is this. A broad, thin sheet of skin is removed from some uninjured part of the patient's body, most conveniently the thigh, laid upon the area from which skin has been lost, and held firmly in position for four or five days until the primary union of the graft to the tissue underneath it is complete. The area from which the graft was removed will heal of its own accord within a week or two, for the graft will not have been cut so thick that epithelium from the inner ends of the truncated hair roots cannot creep upwards and grow over the denuded surface; indeed, one donor area can provide more than one crop of skin.

The use of grafting to make good the loss of skin is satisfactory for wounds up to, perhaps, one square foot in area; but (in the form in which I have described it) it becomes less and less satisfactory as the area of the wound increases, and a severely burnt patient may well have lost far more skin than can be wholly made good from his own resources. The area of loss cannot now be fully covered with grafts; what is done, therefore, is to 'seed' it with small patches of skin in the shape of squares or rectangles, evenly spaced apart. The outgrowth of epithelium from these little skin grafts, combined with ingrowth from the wound margins, forms a new skin surface, and contracture, though it still happens, is reduced in proportion to the area that has been covered with grafted skin. This operation of patch-grafting is avowedly a makeshift; the end-result is neither elegant nor functionally more than serviceable; but it is designed to save a life which might otherwise have been despaired of, and so it does.

What is interesting to the biologist, however, is not what the surgeon does in such a predicament but what he does *not* do. If a patient cannot afford skin of his own for grafting, why not use skin grafts from someone else? There is no surgical obstacle to such a procedure, and voluntary donors are not hard to come by; nor would there be any difficulty in setting up a 'skin bank' to be drawn upon in an emergency, for skin may be stored alive and without deterioration for a good six months at the temperature of liquid air or carbon dioxide snow. The difficulty is that a skin graft from one human being *will not form* a permanent graft upon the body of another. In the first week or so after its transplantation the homograft* (as it is called) behaves just like a graft which has merely been transposed from one part to another of a single individual. The first outward sign that all is not well is a puffiness and inflammation of the grafted skin, leading to a weakening and ulceration of its surface, and finally to abject necrosis followed by a sloughing away. How soon this happens—perhaps after only ten days, perhaps in a month—depends upon a number of variables, some of which I shall mention below. Every now and again a homograft lasts long enough to make a surgeon begin to hope that a natural law is about to be suspended in his favour, but sooner or later (and the general rule is sooner rather than later) the graft withers up and disappears. A human being is resolutely intolerant of skin grafted upon him from other members of his own species; so is a newt, chicken, mouse or cow; nor will even a goldfish accept a scale from any other.[1] The problem of how this comes about, why it should be so, and what can be done about it is the subject of the present article.

2. RULES AND EXCEPTIONS

The idea that homografts of skin are invariably destroyed within a few weeks of their transplantation, so that they are useless for anything except avowedly temporary repair, is now accepted by all well-informed surgeons, though recognition of the truth was slowly and hardly won. I now wish to consider three exceptions to this general rule, two of them predictable and perfectly intelligible, the third of a surprising and entirely unexpected kind.

* 'Homografts' were universally so described when this essay was written, but since then most immunologists have followed the lead of Dr. Peter Gorer in describing them as 'allografts'.

[1] W. H. Hildemann, *Proc. N.Y. Acad. Sci.*, 1957.

The first exception is that skin and other tissues can be exchanged between identical twins on a scale limited only by the exigencies of surgical technique. For the purpose of transplantation, identical twins behave as if they were a single individual—as indeed they once were, for both arose from a single fertilized egg. Skin grafting has just once been used to clear up a problem of disputed parentage.[1] One day the father of two six-year-old boys Pierre and Victor (*'il s'agit ici de pseudonymes'*) had his attention called to a third small boy who was reported to be the very image of his Victor. The third boy, Eric, had been born in the same clinic and on the same night; it seemed possible, therefore that Eric and Victor were twins and that the children had somehow been muddled up. A very careful character-comparison undertaken by Professor Franceschetti made it all but perfectly certain that Eric and Victor were indeed identical twins, and at least likely, therefore, that Pierre really belonged to Eric's putative mother; but this was an hypothesis which she was unwilling to entertain. Blood grouping tests of some refinement now revealed that Pierre could not have been the son of his alleged mother, but without, unfortunately, excluding the possibility that Eric was his alleged mother's son. (Eric's father was dead.) In this predicament, Sir Archibald McIndoe was asked to perform a test by which the case would be agreed to stand or fall: small skin grafts were exchanged on the one hand between Pierre and Victor and on the other hand between Victor and Eric. The skin grafts exchanged between Pierre and Victor were destroyed and sloughed away in a matter of weeks; those exchanged between Victor and Eric survived over the whole period of observation and presumably survive still. These tests can be taken as a proof that Victor and Eric were twins, and under the circumstances, a proof of 99·9 per cent certainty that

[1] A Franceschetti, F. Bamatter and D. Klein, *Bull. Acad. suisse Sci. Méd.*, **4**, p. 433, 1948; A. McIndoe and A. Franceschetti, *Brit. J. plastic Surg.*, **2**, p. 283, 1950.

they were identical twins. (The reason for this pawky reservation will be explained later.) As to Pierre and Victor, the skin grafting test proved nothing except that they were not identical twins; but as blood grouping had shown that Pierre had certainly been allotted to the wrong mother, Eric's mother at last accepted him as her own.

Just as skin grafts can be exchanged between identical twins, so also can they be exchanged between mice and guinea-pigs which, by having been inbred brother to sister for upwards of twenty successive generations, have come to resemble each other in hereditary make-up almost as closely as if they were identical twins. (This is not necessarily true for all animals, for some have mechanisms which prevent their achieving a very high degree of genetic uniformity.) But even here there is a curious exception of some theoretical importance[1]: in some inbred strains of mice, females will not permanently accept skin homografts from males. It is now pretty well certain that this is because agents of the type that cause the breakdown of homografts are caused to be formed by the Y-chromosome, that which is peculiar to males. There is accordingly no reason why females should not accept grafts from other females, nor males from females or other males, as indeed they do.

When members of two different inbred lines of mice are crossed (supposing them already to have achieved and retained a sufficiently high degree of genetic uniformity), their hybrid progeny, forming the so-called F_1 generation, are also uniform, and will accept grafts from one another. They will also accept grafts from their parents and, mutations apart, from their own progeny of the first or any subsequent generation. Excepting only the special case of grafts transplanted from males to females, F_1 hybrids are therefore the 'universal recipients' of a little microcosm of mice comprising the inbred parental strains

[1] E. J. Eichwald and C. R. Silmser, *Transplantation Bull.*, **2**, p. 148, 1955.

and their immediate or later issue. It is clear, then, that genetic uniformity of donors and hosts is not necessary for the successful use of homografts; what is necessary is that the donors should not possess any substances making for incompatibility which are not also present in the hosts. Mice of the first generation of a cross between the members of two highly inbred strains contain representatives of all the hereditary factors possessed by either parental strain; skin grafted from their parents or their progeny cannot therefore come to them, genetically speaking, as a surprise. All this is of great importance experimentally, but it has no bearing at all on practical everyday affairs, for human beings are, genetically, a most diverse assembly and even the most strenuous effort of abstraction cannot liken them to inbred mice.

The second exception to the rule that skin grafts are sure to perish after transplantation from one individual to another is this: it does not apply to embryos. Embryos will accept not merely homografts but grafts from members of quite different species as well, 'heterografts'. The age at which an animal becomes competent to recognize foreign tissue as foreign varies from one species to another. In sheep certainly,[1] and almost certainly in cattle, the power to react upon and reject homografts has already developed two-thirds of the way through pregnancy; in mice and chickens, the transition from the immature or embryonic to the adult mode of response is marked, to a good enough approximation, by birth itself. These variations are perfectly understandable: if we compare one species with another, it soon becomes clear that birth is a movable feast in the calendar of development, for mice are born at a stage of development not much different from that of a human foetus only three months old, and sheep and cattle are much more mature at birth than man.

[1] P. G. Schinkel and K. A. Ferguson, *Australian J. Biol. Sci.*, **6**, p. 533, 1953.

The third exception, the surprising one, turns out on analysis to be a variant of the second. In some animals it is sometimes possible to exchange skin homografts between *non*-identical twins, that is between ordinary litter mates, which resemble each other in hereditary make-up no more closely than ordinary sibs. This dispensation applies to about 90 per cent of twin cattle—I must be understood to mean *non*-identical twins; that tissue homografts can always be exchanged between identical twins has already been conceded—and, so far as our present meagre evidence goes, to all twin chickens; and it can be assumed to be true of a certain very small proportion of twin sheep and a still smaller proportion, surely less than 0·1 per cent, of human twins. (It was for this reason that I put Victor's and Eric's chances of being *identical* twins no higher than 99·9 per cent.) Twin chickens, I should explain, are those that hatch from a two-yolked egg, each yolk being a separate egg as the embryologist understands that word. They could therefore be described with equal propriety as uniovular or binovular: it just depends on what one means by 'egg'.

The non-identical twins between which it is possible to exchange skin homografts are among the most remarkable animals in nature, for they are graft-hybrids or chimeras; each twin is a mixture of cells of two genetic origins, most of its cells being its own, the remainder having been at one time the property of its partner. The exchange of homografts between them in later life does not therefore make them chimeras; it merely makes them more so. Chimerism of natural origin was first described by the American biologist R. D. Owen in 1945; nearly all cattle twins, so he found, contain a mixture, not necessarily a fifty-fifty mixture, of each other's red blood corpuscles. (This is known to be true of non-identical twins in cattle; it can be assumed to be true of identical twins, but it cannot be proved because their red blood corpuscles cannot be told apart.) How does this come about? The origin of the condition was clear enough, for most litter-mates in cattle are of the kind that share a common blood circulation from a very early stage of embryonic life until birth. All blood cells, therefore, and all cells which may circulate in the blood stream on their way elsewhere, can be exchanged between the twins before they are born. But what is specially interesting is that the state of chimerism, of red-cell intermixture, may last for years or perhaps for life—certainly for much longer than the lifetime of a red blood corpuscle, which is not likely to exceed a hundred days. It follows, then, that not merely red cells but the cells which make red cells must have been exchanged in the foetal cross-transfusion; and that, in defiance of the principles formulated in this article, they survived when the animals grew up and continued to manufacture the red corpuscles characteristic of their original owners.

It is for exactly the same reason that twin chickens are red-cell chimeras, for the blood systems of the two separate embryos within the single shell communicate directly with each other. Moreover, my colleagues and I have shown that the state of chimerism can be brought about artificially by joining two eggs together across a vascular bridge, using the method first devised by the Czechoslovak scientist Milan Hašek. Natural chimerism has been described in a small minority of twin sheep, and in one human being, a Mrs McK, who was not known to be twin when, at the age of twenty-five, she was found to contain a mixture of two genetically different types of red blood corpuscle.[1] Her own were of blood O, but they were mixed with corpuscles of group A, the cellular relicts of a male twin who had died when three months old. There is no knowing how long Mrs McK will remain a chimera, but she has now been so for twenty-eight years; probably, in the long run, her twin brother's red blood cells will slowly dis-

[1] I. Dunsford, C. C. Bowley, A. M. Hutchison, J. S. Thompson, R. Sanger and R. R. Race, *British Medical Journal*, **2**, p. 81, 1953.

appear, and so pay back the still outstanding balance of his mortality.

All known chimeras, then, are twins, and all such twins have been found to accept skin homografts from each other for as long as the state of chimerism endures. So when I say that embryos will accept homografts because they are not yet old enough to have learnt the difference between what is native to them and what is foreign, that is only half the story. The other half is that foreign cells introduced into an embryo affect it in such a way that it may never acquire the power to recognize the cells as foreign, and may accept them as its own. It will accept, moreover, not merely the cells which gained access to it as an embryo, but any cell of the same genetic composition that may be transplanted to it in later life. This is the origin of the concept of acquired immunological tolerance, of which I shall say more in Section 5 below.

3. MECHANISMS

Skin homografts are destroyed by an immunological reaction, that is by a process fundamentally akin to that which is provoked by bacterial, viral or cellular infections, or by the injection of foreign proteins or polysaccharides. For all the clinical good will and perhaps even mortal urgency that accompanies their transplantation, skin homografts are treated as if they were a disease of which their destruction is the cure. (This outdoes *Erewhon*: the disease is beneficial, the cure does harm.) Transplantation immunity therefore resembles the allergies and hypersensitivities and serum sickness, and transfusion accidents and haemolytic disease of the newborn, in being an immunological reaction-gone-wrong; and so far have we travelled from the days when all immunological reactions were supposed to be necessarily beneficial (however mysterious their benefactions might appear to be) that immunology is now hardly less commodious than psychology in providing an etiological funkhole for diseases of which the physical causes are still unknown. That *most* immunological reactions do good is of course a truism, though it is a truism whose truth has been formally proved only in very recent years. There exists a congenital affliction, *agammaglobulinaemia*, the victims of which are unable or almost unable to manufacture blood protein of the class, gamma globulins, to which most antibodies belong. Being virtually unable to manufacture antibodies, sufferers from agammaglobulinaemia go down with almost every infectious disease they may be exposed to, and perhaps with the same disease again and again. Only antibiotics can keep them alive, to be witnesses to the truth that, under normal circumstances, immunological reactions are necessary not merely for remaining in health but for remaining alive at all. The recognition of the disease in its severest form had therefore to await the discovery of antibiotics.

But, it may be objected, how can we be certain that transplantation immunity is not a blessing in disguise—that it is not deeply harmful to mix up tissues of different genetic origins in a single individual, the very thing which transplantation immunity normally makes it impossible to do? A few years ago this argument or innuendo might have carried weight, but now it is no longer tenable. Chimeras occur naturally or, as I shall explain later, can be made artificially. Compared with ordinary animals, chimeras are at no disadvantage; or, if they are at a disadvantage (I am thinking of 'freemartinism', the sterility of most female members of twin pairs of unlike sex in cattle), it is for reasons unconnected with chimerism as such. There is therefore no danger that the attempt to make homografts permanently acceptable is going to brush aside some prudent natural safeguard against compounding individuals of cells of different origins. There is only one special circumstance in which we need expect trouble: when the host will accept the graft but the graft will not accept the host.

Skin homografts are destroyed by what is technically known as an 'actively acquired' immunity reaction; there is no ready-made resistance against homografts, in the sense in which an individual of blood group O is already equipped with antibodies capable of agglutinating red blood corpuscles from donors of groups A or B. Resistance to homografts develops in the course of, and as a consequence of, exposure to the foreign substances, *antigens*, contained within them. At first, as I have already said, a skin homograft behaves just like skin merely transposed from one part to another of a single individual: it heals on just as soundly, it is as quickly and as richly re-equipped with a working vasculature, and it undergoes just the same processes of internal reorganization and repair. It may even survive long enough to grow new skin glands and a new crop of hair. But sooner or later, a reaction overtakes it. Just how soon that happens depends upon many variables, e.g. the quantity of foreign tissue that is grafted, for the more that is grafted, the sooner will it be destroyed. There is, however, one variable whose influence overrides all others, the genetical relationship between the donor and the host, and this is worth a moment's notice.

Philosophers make a distinction between differences of degree and of kind, but the inborn differences between individuals cannot be classified in either way. The differences between individuals are combinational, or, as mathematicians say, combinatorial differences; one individual differs from all others not because he has unique endowments but because he has a unique *combination* of endowments. The number of hereditary factors from which these combinations can be built up, though large, is finite, but the combinations themselves are far more numerous than the individuals who can enjoy them, so that for each man actually on stage there are hundreds of possible men still waiting for a cue behind the scenes.

It follows that although the mechanism of heredity may be ultimately atomic, the relationship between human beings is defined by a virtually continuous spectrum of affinities, bounded at one end by identical twins and extending the other end far beyond the genetically visible region into the affinities between animals which are not members of the same species or even of the same order or class.

The technique of skin grafting is particularly well qualified to demonstrate these propositions, for it can reveal (*a*) that all individuals, with the exceptions already noticed, are immunologically unique; (*b*) that the immunological differences between individuals are combinational in character; (*c*) that the combinations are so diverse that there is an almost continuous range of variation in the acceptability of foreign tissues to their hosts. It is bounded at one end by grafts exchanged between identical twins, grafts which survive as long as the animals which bear them; and close to this end lie, for example, the grafts which have been transplanted from males to females of the same inbred line and which may survive their transplantation by as much as fifty days. Grafts transplanted between ordinary members of the same species normally survive for little longer than a week; and when donors and hosts are members of different species, the grafts ('xenografts') never heal in properly, and it is only in a narrowly technical sense that they can be said to survive for any length of time at all. The range therefore extends from something near zero to something as near to permanence as mortality allows, and this variation is an expression, perhaps the completest single expression, of the genetical relationship between the donor and the host.

One thing that must be said immediately about the response of the recipient, for otherwise one or two of the experiments I shall describe below will be unintelligible. It is only when a human being or other animal is confronted with a homograft *for the first time* that the homograft enjoys a latent period during which it behaves like a graft of the recipient's own skin. A second graft from the same donor, transplanted after the rejection of a first, is set upon almost immediately; it does not heal properly, it never acquires a working vasculature, and it never even begins to reorganize itself internally or to develop new skin glands or hair. From a surgical point of view its destruction is virtually instantaneous, though its epithelial cells can survive a few days until they die of inanition. If, however, the second graft comes from a donor genetically different from the first, its behaviour may be almost completely normal. That is what one would expect, unless the donors of the first and second grafts happen to be closely related, in which case the second graft is summarily destroyed. This behaviour strengthens the analogy between an animal's reaction against homografts and its reaction against a disease. When exposed for the first time to a homograft or a disease-producing organism, an animal *takes* the homograft or may *get* the disease. Resistance develops in the course of exposure—the disease is got over, the homograft sloughs away—and now, for many months, the animal becomes refractory, and will not get the same disease or take a graft of the same kind again. As to the specificity of the reaction, here too there is an analogy, for recovery from one disease will not prevent one's succumbing to another unless, as with cowpox and smallpox, the organisms that cause them are closely related; and so it is with homografts, as I have just explained.

It has been known for many years that bacteria which gain a foothold in the skin may enter the lymphatics, the system of vessels responsible for the fluid drainage of the tissues, and so enter the lymph nodes which lie athwart every lymphatic vessel somewhere between its source in the tissues and its final outflow into the veins. Lymph nodes, less properly lymph 'glands', are the organs in the neck and armpit and elsewhere which people refer to when they say their 'glands' are swollen; and lymph nodes are probably the first places in the body in which antibodies are made. Antibodies made in one animal can be injected into another, so conferring upon it a vicarious, 'passive' or second-hand immunity. In medical practice, for example, antibodies against the toxins of tetanus or diphtheria organisms are commonly made in horses. The horse's serum, now an antiserum, can then be injected into human beings.

The reaction against homografts is much the same. Antigenic substances from the homografts reach the local lymph nodes, normally via the lymphatics, and the local lymph nodes are the principal seat of the recipient's reaction. Mitchison[1] has shown that if cells are taken from the lymph nodes of a mouse which has been actively immunized against a homograft, and if these cells are then injected into a second mouse of the same inbred strain, the second mouse behaves exactly as if it had itself been actively immunized beforehand, i.e. as if it had received and rejected a homograft before. This is analogous to passive immunization with an antiserum in the sense that it is the first mouse which undertakes the reaction against the homograft and the second which shares the benefit of it. But it is not *exactly* analogous, because the transference of the state of immunity cannot be brought about by injecting the first mouse's blood or blood serum into the second; it must be the first mouse's living lymph node cells.

This is the first sign of an important difference between the immunity caused by bacteria (or other remotely foreign antigenic substances) and the immunity caused by living cells originating from some other member of their recipient's species.

[1] N. A. Mitchison, *Proc. Roy. Soc. B*, **142**, p. 72, 1954.

There can be no doubt that antibodies are the chief instruments of the defensive reaction against what may be compendiously described as 'germs'. It is true that antibodies themselves do not seem to do bacteria much harm; what they do is to make the bacteria particularly palatable to phagocytes, or to make them sensitive to the action of a complex constituent of the blood known as 'complement', which dissolves them. But although antibodies may not be *sufficient* to bring about the destruction of bacteria, they are certainly necessary; yet in the reaction against skin homografts there is no clear evidence that they are even necessary.

The highly skilled researches of Dr P. A. Gorer of Guy's Hospital have shown that antibodies are certainly formed when a homograft of skin or of tumour cells is reacted upon and sloughed away. The antibodies are of a chemically quite orthodox kind, and may be recognized by their power to agglutinate the red blood corpuscles of the donor. If antibodies are formed, why should we doubt that they are the chief effectors of the immunological response?

The main reason, perhaps, is that a state of immunity cannot be transferred from one mouse to another by transfusions of blood or serum, for all that the blood may contain a high concentration of antibodies. A second is that the serum of an animal which has received and rejected a homograft contains nothing that opposes or even discourages the growth of a donor's cells in tissue culture. A third has emerged from some ingenious work carried out at the National Cancer Institute in the U.S.A.[1] If a donor's cells are grafted into an animal which has been forearmed against them, by having received and rejected cells of the same origin before, they are promptly set upon and destroyed. But if the donor's cells are housed inside a little permeable plastic bag within the recipient, then

[1] J. M. Weaver, G. H. Algire and R. T. Prehn, *J. National Cancer Inst.*, **15**, p. 1737, 1955.

they will be destroyed if, and only if, the walls of the bag are permeable enough to let through the recipient's cells. Permeability to molecules of the size of antibody molecules is not enough. Clearly, then, it is not sufficient merely to confront a donor's cells with antibodies. Is it even necessary to do so? Apparently not, for the following reason. A donor's tissue, lying in a plastic bag that is permeable to antibodies but not to cells, will be destroyed if cells from the immunized recipient are added to the contents of the bag before it is sealed up and introduced into the recipient. This experiment reproduces the state of affairs in which the donor's tissue is housed in a plastic bag that is permeable to the recipient's cells, the only difference being that, in this variant, the recipient's cells are spared the exertion of getting in. But the donor's tissue is also destroyed if the bag and its contents are grafted into the *donor* or an animal genetically similar to the donor. The donor's body fluids may well contain something necessary for the immunological reaction to take effect within the plastic bag, but whatever that may be, it cannot be an antibody, for the donor cannot contain antibodies acting against its own cells.

For these and other reasons, most of us are now convinced that antibodies play no necessary part in the reaction that destroys skin homografts and most other homografts of solid tissues. With homografts of isolated cells, and more particularly of leukaemic tumours, it may be a different story. By 'antibodies' I must be assumed to mean the ordinary, orthodox antibodies that circulate freely in solution in the blood stream and can be recovered in serum free from cells. Perhaps the cells that do destroy homografts are transporters of antibodies and perhaps they liberate them exactly where they can do most damage, that is, in the immediate neighbourhood of the grafts; but if that proves to be the case, I suspect that the antibodies will turn out to be so far different from ordinary antibodies as to deserve some different name.

4. WHAT CAN BE DONE ABOUT IT?

It is at present possible to envisage four kinds of ways in which a homograft could be made acceptable to its host: (*a*) the antigenic constitution of the graft might be changed, so that it no longer stirred up a reaction in its host; (*b*) the graft might be transplanted in such a way that it could not exercise its antigenic properties; (*c*) the host might be changed in such a way that its reaction against the graft was enfeebled or done away with altogether; or (*d*) the graft might be put in a position in which, no matter what state of immunity might prevail, it could not be got at by the cells that put the immunity into effect.

The first solution cannot be applied to homografts of the kind I have been particularly concerned with, grafts which are alive when they are grafted and which must remain alive if they are to do their recipient any permanent good. Why this should be so has already been explained: the antigenic make-up of a graft is built into its genetic constitution. I therefore grieve at the theoretically infirm attempts which have been made to change the antigenic constitution of a graft by, for example, growing it as a tissue culture in the body fluids of the animal or human being on which it is ultimately to be transplanted—accustoming it gradually (such is the feeble hope) to what it will have to make do with later. No antigenic transformation is in the least likely to occur under these conditions. Antigenic transformations *can* occur under very special conditions which have no bearing on the way in which homografts are used in surgical practice; for example, if a graft consisting of isolated tumour cells is transplanted to an animal which puts up a certain feeble resistance to its growth, then the population of tumour cells, considered as a whole, may change its antigenic properties; but that I conceive to be due to a process of natural selection, i.e. the selection, from a rapidly growing and probably variable population, of the particular variants that are least antigenic to the host.

Not all grafts are of the kind that need remain alive after their transplantation; homografts of segments of blood-vessels, for example, are building up an impressive record of successful use in human surgery, but, to put it as a paradox, they are successful as homografts because their failure does not matter. What a vascular homograft does is to provide a fibrous tube of the right shape and texture which, when its own cells die, is repopulated by cells arising from its recipient; a new lining of endothelial cells is laid down on the inner surface of the graft, and cells of the connective tissue family penetrate the interstices of its fibrous skeleton and convert it into a plausible and efficient imitation of a normal blood vessel. That vascular homografts, as living entities, 'die' and are none the worse for it is shown by the fact that they need not be alive even to begin with. Vascular homografts which have been killed (though kept in a lifelike condition) by drying from the frozen state or by prolonged storage at very low temperatures seem to do as well as living grafts; synthetic plastic tubes will also serve. Vascular homografts die, but they enjoy the privilege of reincarnation.

The second and third ways of getting round the homograft reaction are complementary to each other; either to transplant a homograft into a position in which the antigens manufactured by it never reach a centre of response, so that the host is never officially aware of its existence, or to transplant it into a position in which the effectors of the immunological response are unable to get at it. The main seat of the host's reaction against homografts, I have explained, is the regional lymph nodes, and antigens reach them through the regional lymphatics. The brain has no lymphatic drainage in the ordinary sense, nor is it monitored by lymph nodes. It is therefore entirely intelligible that homografts transplanted into the substance of the brain should often fail to elicit an immunological reaction.

The complementary case is best exemplified by the cornea. Homografts transplanted into the cornea are in a kind of sanctuary. Their condition is to be likened, perhaps, to that of homografts transplanted into the little plastic bags, permeable to fluids but not to cells, which I mentioned in an earlier section. The cornea is a non-vascular structure, so that the cells which are the effectors of the host's reaction against homografts simply cannot get through it—unless, indeed, the cornea should become vascularized by accident, in which case a graft transplanted to the cornea will usually fail. Brain and cornea are therefore both privileged sites of grafting, but for entirely different reasons. A simple experiment will make the distinction clear. A homograft in the brain will certainly be destroyed if the animal into which it is transplanted is immunized by some other, efficacious route—for example, by a skin homograft transplanted upon or beneath the skin. It is therefore entirely vulnerable to an immunological reaction; it owes its privilege only to the fact that it cannot set such a reaction going. But a homograft in the cornea, provided the cornea is unvascularized, will survive in the face of a fulminating immunity directed against it; it survives because the state of immunity cannot take effect.

So much for what may be called the bye-laws of tissue transplantation, the special rules that govern the behaviour of homografts in special positions in the body. I have mentioned two or three, but others remain to be discovered. No one has yet put forward a plausible explanation of why it is that homografts of certain endocrine glands—of the ovary, for example, in so far as it is a source of hormones, or the adrenal cortex—sometimes survive when homografts of skin demonstrably fail to do so. But a *general* solution of the homograft problem must turn upon the last of the four expedients which I mentioned in the introduction to this section, i.e. upon changing the host in such a way that its reaction against homografts is done away with or at least enfeebled.

There are half a dozen ways in which the intended recipient of a homograft can be treated in order to prolong the homograft's normal lease of life; most are temporary, but one can be permanent; some could be applied in surgical practice if it were worthwhile doing so, others not; a few are innocuous, but most are harmful. By far the most important distinction, however, is between treatments which are non-specific and treatments which are specific in their action. By a non-specific treatment I mean a treatment which will weaken, or under extreme circumstances abolish, the reaction against homografts from all sources—and, for full measure, probably abolish most other immunological responses as well. A specific treatment is one which weakens the reaction against homografts from some one particular donor, or from the members of some one highly inbred strain, without prejudice to the reaction against homografts from other sources or, *a fortiori*, to the recation against antigens of other quite different kinds.

I shall mention only two of the non-specific treatments: X-irradiation, and the injection of cortisone; and I mention X-irradiation simply because it has the awful prestige of anything to do with the threat of atomic war. 'Whole body irradiation' of a sufficient dosage—for a mouse, something less than one thousand Roentgen units—causes, amongst other things, complete immunological prostration and severe and usually irreparable damage to blood-forming cells. It was discovered in America that mice which had received a dosage of radiation which would otherwise have been rapidly fatal could be kept alive by injecting them with cellular pulps made from the blood-forming tissues of other mice or even of members of alien species. For many years the nature of the protective agent contained within this pulp remained in doubt; some maintained that it was humoral in nature, and that it could exist apart from living cells, others that the injection of the pulp was in effect a transplantation of normal living cells which simply took the place of those damaged or destroyed by radiation. Two independent groups of scientists,[1] one at Harwell and the other at Oak Ridge, have shown that the latter explanation is certainly correct. A heavily irradiated mouse is no longer capable of resisting the transplantation of foreign, even of remotely foreign, tissue; when it is injected with blood-forming cells from bone marrow or other blood-forming centres of normal mice, the cells establish themselves without opposition in their new surroundings, and the irradiated mouse becomes a 'radiation chimera' in which the foreign blood-forming tissues act proxy for its own.

The injection of high doses of cortisone, a drug which is closely related to one of the natural secretions of the cortex of the adrenal gland, produces an effect which has something in common with radiation sickness. Steroid hormones of the class to which cortisone belongs have a powerfully inhibitory effect upon the growth and activity of all lymphoidal tissue—upon, therefore, the cells which undertake their owners' immunological reactions. The injection of large doses of cortisone can certainly prolong the life of homografts, though (for reasons which are still not quite fully explained) it does so more readily in mice and rabbits than in guinea-pigs or human beings. At one time we hoped that cortisone would be a useful minor addition to the armoury of the plastic surgeon in the treatment of very extensive burns. The great raw wounds left by deep and widespread burns are still sometimes covered with homografts of skin; homografts make a perfect temporary dressing which

[1] C. E. Ford, J. L. Hamerton, D. W. H. Barnes and J. F. Loutit, *Nature*, **177**, p. 452, 1956; D. L. Lindsley, T. T. Odell and F. G. Tausche, *Proc. Soc. exp. Biol. Med.*, **90**, p. 512, 1955.

may tide the patient over until he can afford to provide some skin grafts of his own. It would buy useful time if these homografts were made to survive only twice as long as could otherwise be expected, and this is what we hoped that cortisone would do. Unfortunately, it seems that cortisone could prolong the life of homografts on human beings only at dosages which would have secondary ill effects of a gravity which an already sick patient could not put up with. Yet the research account is by no means all debit, for study of the action of cortisone and other steroid hormones on the behaviour of homografts, is giving us a new insight into the nature of the normal adrenal cortical secretions, how they vary from members of one species to another, and how they are influenced by the trophic hormone of the pituitary gland.

The weakening of an animal's reaction against homografts which can be brought about by the treatments or maltreatments I have just described is simply a by-product of some more general biological damage. The specific treatments I now turn to are nicely discriminating in their action; they influence the survival of homografts from particular donors chosen beforehand, and have no effect on other immunological reactions. One such treatment, the particular study of research workers in the Roscoe B. Jackson laboratory at Bar Harbor, Maine, entails the injection into animals which are later to receive homografts of desiccates or extracts made from the tissues of their future donors. The theoretical importance of this treatment outweighs its practical usefulness, which by all appearances is very modest, for the best it has been able to do in our hands is to double or treble the normal expectation of life of a skin homograft. Beyond the fact that it is certainly immunological in character, there is no common agreement about the way this treatment works.

The second 'specific' method of interfering with the homograft reaction is that which turns upon the principle of immunological tolerance, and it deserves—or at all events is to receive—a chapter to itself.

5. IMMUNOLOGICAL TOLERANCE

When antigens are injected into juvenile or adult animals, they provoke some kind of immunological response. That is not an empirical fact but a tautology: 'antigens' are so defined. It is an empirical fact, however, that when antigens are injected into embryos, or into newborns of the kind that are born very immature, they do not elicit an immunological reaction. For many years immunologists were content to dismiss this fact by saying that the immunological faculty is one that develops and matures like any other, and that embryos do not react upon antigens simply because they are not yet sufficiently grown up. This is a half truth; the other half of the truth is the subject of the present chapter.

In 1949, F. M. Burnet and Frank Fenner propounded a theory of antibody formation which led them to make the following prediction: that if an embryo were to be injected with an antigenic substance, then, when it grew up, its power to react against that antigen would be found to have been seriously impaired. Almost all they had in the way of hard facts to go on was Owen's discovery (p. 130) of red-cell chimerism in twin cattle—the state of affairs in which cattle twins are born with and long retain a mixture of each other's red blood corpuscles, presumably because they exchange blood-forming cells in embryonic life. If the exchange had not occurred before birth, but had been carried out artificially afterwards, then the foreign blood-forming cells would quite certainly have been recognized as such and destroyed by an immunological reaction. The exchange of the cells before birth must somehow have prevented the development of that faculty which would have empowered the twins to recognize each other's cells as not their own.

My colleagues and I have shown that Burnet and Fenner's prediction is true, without qualification, of the antigens which are responsible for transplantation immunity. We too began our work on cattle. In 1948, while attending an International Congress of Genetics at Stockholm, I was invited by Dr H. P. Donald to help to solve the important problem of distinguishing with complete certainty between identical and non-identical twins in cattle. In principle, nothing could be easier. Skin grafts were to be exchanged between the twins a few weeks after their birth. If the homografts survived, the twins could be classified as identical; if not, as non-identical, i.e. dizygotic. My colleague R. E. Billingham and I, helped by two young officers of the Agricultural Research Council, began what was to be a few months' work later on that year; as it turned out, the work took three years to finish. We satisfied ourselves that the reaction against skin homografts was no less vigorous in cattle than in the man or in laboratory animals, and that skin grafts exchanged between cattle of the same or of different breeds, or between dam and calf or vice versa, or between ordinary siblings (brothers or sisters but not twins) were all destroyed within a fortnight of their transplantation. But homografts between almost all the twins survived, irrespective of whether the twins were identical or dizygotic. There could be no mistake about the classification of the twins as dizygotic, for the pairs we chose for our critical tests were of unlike sex and as different as possible in other ways. Dizygotic twin calves were therefore tolerant of homografts of each other's skin—though not, it should be observed, of the skin of any other cattle. Obviously the cross-transfusion in foetal life which led to their becoming red-cell chimeras had destroyed their power to recognize each other's skin as foreign. M. Simonsen later reported the successful exchange of whole kidneys between dizygotic cattle, so it is likely, as more recent experimental work has shown to be true of mice and other laboratory animals, that dizygotic twin cattle are tolerant of homografts of *all* tissues from their twins, though of no tissue from any other animal.

Billingham, Brent and I therefore set ourselves to reproduce experimentally, in laboratory animals, the state of affairs that occurs by a felicitous natural accident in twin cattle and in those other natural chimeras mentioned in Section 2; and after a year's labour we succeeded.[1] A typical experiment was conducted thus. The seventeen-day old embryos of white mice of strain A were injected, while still *in utero*, with a mixture of cells taken from an adult donor belonging to the quite different, brown, strain of mice, strain CBA. The injected mice were born a few days after their injection, and allowed to grow up. A normal adult mouse of strain A rejects skin homografts from CBA mice within eleven days of their transplantation, but adult mice which had been injected before birth with CBA cells were, in the extreme case, completely tolerant of grafts transplanted from CBA donors. Graft hybrids could therefore be made at will. Since these first experiments were done, the technique has been greatly simplified—with some strains of mice, for example, the preparatory injection can be delayed until immediately after birth—and has been extended to other mammals and to birds.

It follows, then, that the 'homograft reaction' can be completly abrogated; the problem of making animals completely tolerant of foreign tissue is not, as I had once feared, insoluble. But it must be said at once that it is hardly feasible to apply this technique to human beings—not so much because it would require interference with an unborn baby, though that objection is grave enough, as because the state of tolerance is absolutely specific. The white mice I referred to above, made tolerant of CBA grafts by injecting them before birth with CBA

[1] A full account of this work is contained in R. E. Billingham, L. Brent and P. B. Medawar's monograph in *Philos. Trans, Roy. Soc. B*, **239**, p. 357, 1956.

cells, invariably reject skin homografts from other, unrelated donors; so likewise a human foetus injected with (for example) blood cells from Mr Smith (il s'agit ici de pseudonymes) could not be expected to accept homografts in later life from anyone except Mr Smith himself.

Among the multitude of experiments which my colleagues and I have done in course of our analysis of the phenomenon of tolerance, two only will be singled out for special mention. The first is that the state of tolerance does not discriminate between the different tissues of a single individual; an injection of blood into the embryo will cause a tolerance of skin, an injection of spleen cells will cause tolerance of a graft of the cortex of the adrenal gland, and so on. If a tissue A, injected into an embryo, is to cause tolerance of some other tissue B, grafted later in life, then clearly B must contain no antigen that is not also present in A. (If B had some antigen peculiar to itself, there is no reason why it should not go into action and immunize the host.) This condition appears to be fully satisfied when A and B are any two different tissues from the same individual. This is of great practical importance.

The second property of tolerance I wish to mention also bears directly upon the nature of the antigens that cause the homograft reaction. A state of tolerance, no matter how long it has prevailed, can be brought to an end simply by re-equipping the tolerant animal with normal, and therefore immunologically competent, lymph node cells. Consider an A-line mouse which has been made to accept a homograft of skin from a mouse of strain CBA, and let it be supposed that the homograft, bearing its characteristic coat of brown fur, has long been fully accepted by its host. The homograft can be destroyed, and the state of tolerance brought permanently to an end, simply by injecting the A-line mouse with normal lymph node cells taken from normal A-line donors. Until then, the antigens given forth by the CBA skin homograft were unable to elicit a reaction, because its recipient's lymph node cells had been incapacitated by the CBA cells to which they were exposed in embryonic life. But when the A-line host has been refurnished with normal lymph node cells, then the antigens liberated by the graft have a chance to act upon lymphoid cells which are in normal working order; immunity is built up and the homograft is destroyed. The peculiar importance of this experiment is that it shows that the CBA skin homograft, though a perfectly normal tissue, *was manufacturing antigens all the time*—or rather, was manufacturing substances that would have been antigenic if only the host had been competent to recognize them for what they were.

The problem of how tolerance comes about is still unsolved. For the present, the phenomenon of tolerance must be accepted as one of the raw data of immunology, and no theory of the mechanism of the immunological reaction will pass muster unless it can explain the phenomenon of tolerance as well. J. B. S. Haldane suggested that an embryo has the power to metabolize—to break down and make use or dispose of—substances which the adult cannot metabolize; the adult makes antibodies against them instead. If that is so, then tolerance is the enforced retention of an embryonic modality of response; and it would be in keeping with Burnet and Fenner's theory if the transformation which leads to tolerance were closely akin to the 'training' of bacteria to metabolize unaccustomed foodstuffs or to resist the action of inhibitory drugs. I think this interpretation is plausible, because to secure a state of tolerance it is probably not sufficient merely to confront an embryo with antigens; the antigens must persist, and continue their educative action, well into the period in which, under normal circumstances, the young animal would have become immunologically mature. Fortunately, we need not commit ourselves to any particular theory of the mechanism of tolerance before examining its biological implications, which are various, though I can consider only one or two.

The phenomenon of tolerance requires one to think anew about the nature of the relationship between a mammalian mother and her unborn young. Except in highly inbred animals, a foetus has a different genetic and therefore a different anti genic constitution from its mother. It is therefore an antigenically foreign body, a kind of foreign graft. Why then does it not immunize the mother, with consequences disastrous to itself?

Haemolytic disease of the newborn is evidence that this does happen sometimes; that it happens very seldom is due to the extraordinarily efficient insulation of the mother from her unborn young. Under normal circumstances, no particles as large as cells, could possibly cross the placental barrier from the foetus into the mother's blood. This arrangement provides against the danger that the foetus should immunize the mother, but this is not the only immunological danger, and a one-way control of traffic between foetus and mother is not enough. If particles as large as cells could pass from the mother to the foetus, then infective organisms or the antigens manufactured by them could also do so, and although maternal antibodies might keep infection in check, that would not prevent the antigens of micro-organisms from damaging, perhaps irreparably, the future development of the immunological defences of the child. Nearly thirty years ago Traub showed that a virus disease of mice, lymphocytic choriomeningitis, can be transmitted from a mother to her unborn young; the young were accordingly unable to develop resistance to the virus in later life, and might transmit it to their young in turn. In this particular strain of mice, virus and host had come to a live-and-let-live arrangement by which neither killed the other, and reflection will show that, had this not been the case, the phenomenon could hardly have been discovered. But here is an example of a 'hereditary' infectious disease, running in a family, but transmitted from mother to young because each generation not merely infects its successor but abolishes its successor's power to rid itself of the disease. 'Genetic predisposition' is therefore not the only possible explanation of a tendency of certain diseases to run in families

Under normal circumstances, the mere incorporation by a foetus of some of its mother's cells need not be expected to lead to evil consequences; it does not normally happen because, as I explained above, a frontier which lets through cells would let through undesirable immigrants as well. One can be confident that maternal cells are *not* admitted into the foetus, because if they were, the young should acquire a complete or partial tolerance of homografts transplanted from the mother, and this happens very rarely, if at all. Cancerous cells, however, are distinguished by their invasive properties, and there are half a dozen cases on record of the apparent transmission of malignant melanomatosis from a pregnant mother to her child. There can be little doubt that the melanoma cells actually crossed the placental frontier and established themselves in the foetus. An adult human being certainly, and even I think a newborn, will destroy homografts of malignant cells. It would not inevitably be disastrous for a foetus to allow, because it could not prevent, the growth of a foreign malignant tumour, provided only that it could rid itself of the tumour as soon as it became immunologically mature. Unfortunately, the effect of exposing the foetus to the malignant cells would be to prevent that very process of maturation, so that death early in post-natal life would be almost inevitable.

When therefore we think of the immunological relationship between the mother and the foetus, we must read the relationship both ways round: the foetus must not be allowed to immunize the mother, and the mother must not be allowed to weaken the immunological defences of the child. It is for this reason, and for no other one sufficient reason, that the blood systems of the mother and the foetus must be strictly separate all the time, in every place, and at every level down to the finest capillary vessel.

Beyond this, the concept of immunological tolerance has implications which are deeply philosophical, in the worst sense of the word, for it bears directly upon the problem of the recognition and awareness of The Self. Why do not the cells which undertake immunological responses react against constituents of the body in which they themselves are housed? Why are not 'auto-antibodies' regularly formed? Alas for *Naturphilosophie*, the problem is soluble and can be clearly put. The question of manufacturing auto-antibodies (or otherwise reacting) against antigens of the type that cause transplantation immunity cannot arise in practice, because, as I have already explained, these antigens are uniformly represented in all the tissues of a single individual, not excepting his antibody-forming cells. Being part of the fabric of his own antibody-forming cells, an individual's own 'transplantation antigens' cannot be reacted upon as if they were foreign. But that does not explain why, for example, muscle protein, or any proteins distinctive of skin or nerve, should not appear foreign to an individual's own antibody-forming cells. The answer, we believe, is that his antibody-forming cells develop in the constant presence of, grow up with, these very substances, and so, in the technical sense I have just explained, become *tolerant* of what might otherwise have been their antigenic action. This explanation sounds too facile to be true, but fortunately there are certain exceptions which seem to prove the rule. Antibody-forming cells obviously cannot become tolerant of any bodily constituents which are formed, or do not become mature, until the antibody-forming cells themselves have formed and become mature; for example, they could not become tolerant of milk protein or chemically distinctive ingredients of spermatozoa. Nor could antibody-forming cells become tolerant of bodily constituents which, however early they develop, are physiologically shut off from the remainder of the body, e.g. by lacking a blood supply or lymphatic drainage; they should not therefore become tolerant of the potentially antigenic action of the characteristic proteins of the lens. If this interpretation is correct, then substances like milk protein and lens protein and spermatozoa should be capable of forming auto-antibodies, though needless to say they never get a chance to do so in ordinary life. So they are: appropriately administered, all can form auto-antibodies in the body of which they themselves are part. The phenomenon of tolerance is therefore of fundamental importance in the mechanism by which the body learns to discriminate between what is proper to itself and what is foreign, and it is only under artificial or otherwise abnormal circumstances that the mechanism of recognition goes wrong.

6. CONCLUSION

In this article I have shown how skin grafting can be used for the detection and assay of individuality, whether in goldfish, mice or men. Although the inborn differences between human beings are combinational in origin and inner structure (they are not to be thought of as differences of either 'degree' or 'kind'), yet the combinants are so numerous, and so generous are the ways in which they may be combined, that every human being is genetically unique; the texture of human diversity is almost infinitely close woven. But what is the 'meaning' of this diversity, i.e. what intelligible function does it fulfil? That is not a question one can very well ask of human beings, because the answer would be too complicated and too hedged around with qualifying clauses; but the gist of the answer, as it relates to lower organisms, is this. Inborn diversity makes for versatility in evolution. Every living species must provide not only for the present but also for what may happen to it in the future; only those lineages survive to the present day which, in the past, were versatile enough to come to terms with their environment. All organisms must have a genetical system, as they must also have immunological and nervous systems, which can cope efficiently with what has not yet been experienced—with what, if they were sentient, we should call the unforeseen. Bacteria and other micro-organisms, for example, must have a genetical system which will protect them as effectively from antibiotics which have yet to be discovered as from those which they have coped with hitherto. Only inborn diversity, and a genetical system which keeps that diversity permanently in being, can make this possible. It is a mere truism that if inborn diversity and genetic individuality were to be extinguished, as in some animals they can be, by inbreeding, then selection would have nothing to act on, and the species would be left without evolutionary resource. Curiously enough, this would probably be less harmful to human beings than to any other animal, for men have devices for avoiding the rigours of selection, and can change the environment instead of letting the environment change them. So far from being one of his higher or nobler qualities, his individuality shows man nearer kin to mice and goldfish than to the angels; it is not his individuality but only his awareness of it that sets man apart.

Index

Abercrombie, M. 70
acquired characters. *See* Lamarckism
adaptation 69–73, 75
—bacterial 63–6, 85–6
agammaglobulinaemia 133
Agar, W. E. 79
age-distribution 21, 43–4
ageing 30. *See also* senescence
Algire, G. H. 138
Allee, W. C. 6
allergy 103
Allison, A. C. 106
allografts xvii, 126
anabiosis 13
Andjus, R. 14
antibiotics 63–6, 124, 133, 155
antibodies, *See* immunity reactions
antigens. *See* immunity reactions
antiserum and eye defects 76–7
Aschoff, L. 2

Bailey, W. T. 86
Baldwin, J. M. 66
Baly, W. 15
Bamatter, F. 127
Barnes, D. W. H. 144
Basques 107
Bateson, W. vii
Beadle, O. 108
Beale, G. H. 84
Beet, E. A. 106
behaviour 114–8
—human 114
—rats 78–80
Bell, J. 50
Berkeley, G. 62
Bernard, C. 1
Bernheimer, A. W. 83
Bidder, G. P. 9, 15–6, 41
Billingham, R. E. xv, xvi, xix, 13–4, 68, 70, 110–1, 112, 147–8
blood groups 73, 103–7, 128, 131–2
blood vessel grafts 141
bone 71, 90, 109
Borodin, N.A. 14
Bowley, C. C. 131
brain, mechanical 59, 120
Brambell, F. W. R. 77
Brent, L. xix, 148
Briggs, R. 13
Brown, G. W. 43
Buffon, G. L. L. 1, 8
Burnet, F. M. 146–7, 150
burns 123–5, 144
Burrows, M. T. 10
Burt, W. H. 17
Bushnell, L. D. 76

Cain, A. viii
Cannon, H. G. x
carp 13
Carrel, A. 10

Carr-Saunders, A. M. 76
Champy, C. 101
Child, C. M. 13
chimera 130–3, 144, 146–7
Chitty, D. 17
chorea, Huntington's 50–1
Clarke, C. viii
Clark, W. E. le G. 94
Coale, A. J. xiii
Comfort A. 6, 12
cornea 70, 142–3
corns 69–71
cortisone 143–5
Cowdry, E. V. 11
Crew, F. A. E. 79
cytoplasmic inheritance 84–6

Darlington, C. D. xviii
Darwin, C. 68
Darwinism vii, ix, 63, 66, 73–4
—and Lamarckism 63–87
death. *See* senescence
—'natural' 2, 39–41
de Beer, G. R. 72
Deevey, E. S. 6
Delbrück, M. 86
Descartes, R. 60
diagnosis, clinical 58
differentiation 8, 101
diffusion 93
disks, intervertebral 95, 108–10
Dobzhansky, T. viii
domestication 17–8, 40, 47, 95
Donald, H. P. 147
Dowdeswell, W. H. 6
Drew, J. S. 79
Drummond, F. H. 79
Dunsford, I. 131

Eichwald, E. J. 128
Elton, C. 6
embryo. *See* foetus
Emerson, A. E. 6
endosomatic instruments, evolution 119–22
epidermis. *See* skin
eugenics 24
evolution (*see also* Darwinism, Lamarckism) 18–9, 102–3
—endosomatic, exosomatic xviii, 120–1
experiment, defined 61–2

Fenner, F. 146–7, 150
Ferguson, K. A. 129
fertility 21–3, 38, 45–7, 133
Finlay, G. F. 76
fish, senescence in 15–6, 41
—sizes of 91–2, 97
Fisher, R. A. viii, xii, 6, 21, 23, 53
flexure lines 71, 73
Flood, M. M. 43
Flourens, M. J. P. 8
Flower, S. S. 7, 16
foetus 76–8, 103–5, 129, 150–3
force of mortality. *See* life table
Ford, C. E. 144
Ford, E. B. 6, 80
form (*see also* transformations) 89, 97–101
Fowler, E. H. 83
Franceschetti, A. 127
freemartinism 133
freezing 13, 125–6

Geill, T. 10
genes 12, 22–3, 47–54, 76, 84–5, 106

Good, R. A. xxiii
Gorer, P. A. xx, 47, 138
graft hybrids. *See* tolerance
grafting xxiii, 10, 25–6, 68–9, 73, 79, 110–55
growth 88–101
– differential 38, 49, 94–101
– intussusceptive 111–2
– laws of 94
– rate of 4–5, 31, 88–101
Grüneberg, H. 47
Gunson, M. M. 79
Guyer, M. F. 76–7, 79

haemolytic disease 103-7, 132, 151
Haldane, J. B. S. viii, xii, 12, 51, 76, 79, 93, 107, 150
Hamerton, J. L. 144
Harrison, J. A. 83
Harrison, J. W. H. 80
Harrison, R. G. 10
Hašek, M. 77
Hayflick, L. xiv
Helmholtz, H. L. F. von 102
Hemmings, W. A. 77
Henderson, M. 77
heterografts 129
Hewer, H. R. 81
Hildemann, W. H. 126
Hinshelwood, C. N. xi, 66, 85
Hinton, M. A. C. 17
Holliday R. xii
homografts. *See* grafing
Hughes, A. W. McK. 80
Hunter, J. 13
Huntington, J. 50–1
Huntington's chorea 50–1
Hurst, A. 9
Hutchison, A. M. 131
Huxley, J. S. 49, 76, 98
Hyde, R. R. 77
hypothermia 14
hypothesis 56–8, 60–1

Ibsen, H. L. 76
Ilitch, I. xviii
immortality 10–14
immunity reactions 26, 76–80, 83, 103–4, 132–55
induction, logical 57
instinct 116–8
intervertebral disks 95, 108–10
Irwin, M. R. 76
isotopes 88

Jackson, C. H. N. 6
Jennings, H. S. 11
Jensen, C. O. 10
joints, 71–3, 108
Jund, L. 13

Kermack, K. A. 30
Kimball, R. F. 83, 85
Kirkwood, T. B. L. xii
Klein, D. 127
Kneale, W. 60
Koga, Y. 7
Korenchevsky, V. 7
Krohn, P. L. xvi, 25
Kukenthal, W. 72

Lack, D. ix, 6, 40
Lamarck, J. B. 63
Lamarckism x, xi, 63–87
– defined, 67, 75
– in micro-organisms 83–6
Landsteiner, K. 103
Lankester, E. R. 16–7, 41
Lansing, A. I. xiv

Lederberg, J. 86
Lemche, H. 80
Leslie, P. H. 5
Levine, P. 103, 107
life, expectation of 16–8, 26–9, 33, 51
life table 5–6, 17–21, 32–5, 42–8
Lindsley, D. L. 144
Little, C. C. xx
Lloyd Morgan, C. 66
Loeb, L. 10, 79
Lorenz, K. xvii, 105
Lotka, A. J. viii, 5–6, 26, 119
Loutit, J. F. 144
Lovelock, J. E. 14
lymph nodes 137, 141, 149–50

MacBride, E. W., x, 9
McCay, C. M. 12
McDougall, W. x, 78–82
McIndoe, A. 127
McMichael, Sir J. xxi
malaria 106
mammals, pregnancy 151–3
—senescence in 8–9, 16–9, 40
—sizes of 90–1
Matthews, L. H. 92
Mayr, E. viii
Medawar, P. B. 7–8, 70, 72, 94
melanism 80–1
melanocytes 70
memory 38
Mendel, G. vii, 11
Metalnikov, S. 3
metatheory 58
Metchnikoff, E. 1, 8, 9
method, scientific 55–62
Miller, J. F. A. P. xxiii
Minot, C. xv, 4–5, 7–8, 15, 31–2, 96
Mitchell, P. C. 8
Mitchison, N. A., 104, 137
Morant, G. M. 7, 95
Morgan, L. 66
Morgan, T. H. xviii
mortality, force of 5–6, 17, 20–3, 33–5
Muller, H. J. 75
Müller, J. 15

Needham, J. 13
Neel, J. V. 106
Norton, H. T. J. viii
Noüy, L. du 10

Odell, T. T. 144
Olsen, E. 10
Orla-Jensen, S. 10
Owen, R. D. 146

Palmer, J. F. 13
paramecium 83–5
Park, O. 6
Park, T. 6
Parkes, A. S. 14
Pauling, L. 106
Pearl, R. 12, 14
Pearson, K. 9, 19
penicillin. *See* antibiotics
Penrose, L. S. 51
Pierson, B. F. 12
placenta. *See* foetus
polymorphism xvi, xx, xxi, 73, 105–7
Poulton, E. B. 2
pregnancy. *See* foetus
Prehn, R. T. 13
protozoa, immortality of 11
Pyke, D. A. xii, xiii

Race, R. R. 131
Ranson, R. M. 5
rats, growth and longevity 13
—inheritance in 78–80
reproductive value 21, 23, 44–5, 53
Rhine, J. B. 78–9
Robertson, A. viii
Rowlands, H. A. 59
Russell, B. A. W. 58
Russell, P. S. 112

Sanger, R. 131
Schinkel, P. G. 129
Schmidt, K. P. 6
Schmorl, G. 109
Schönland, S. 2
scientific method 55–62
selection (*see also* Darwinism) 63- 6, 78–80, 85, 140
senescence 1–27, 28–54, 109
—definition of 6, 39
—evolution of 1–27, 28–54
—measurement 31–9
Sheppard, P. viii
Shipley, A. E. 2
sickle cell disease, trait 105–6
Silmser, C. R. 123
Simonsen, M. 147
Simpson, G. G. 18
size 15–6, 90–7
skin, ageing in 36–7
—grafting 123–55
—healing of 110–3, 123–6
—varieties of 68–71
Sladden, D. 81
Smith, A. U. 14
Smith, F. A. 76, 77, 79
Smith, J. M. 92
Snell, G. D. xx
Sonneborn, T. M. 12, 83
Spencer, H. 63, 92, 93
Spencer's Law 92–3
spinal column 107–10
squatting facets 72
stick insects, inheritance in 81–2
Strangeways, T. 11
Sturtevant, A. H. 76–7

Tanner, J. M. 95
Tatum, E. L. 86
Tausche, F. G. 144
Thompson, D'A. W. viii, 98, 100–1
Thompson, J. S. 131
Thomsen, M. 80
Thorpe, W. H. 82
Tiegs, O. W. 79
time, biological 13
Tinbergen, N. xvii, 115
tolerance, immunological xix, xx, 132, 141–54
tradition 114–22
training of bacteria 63–6, 85–6
—of rats 78–80
transformations, Thompsonian 100–1
transplantation. *See* grafting
Traub, E. 151
Trotter, W. 120
twins xiii, 127–32, 135, 147–8

vertebral column 107–10

Waddington, C. H. 70, 74
Wallace, A. R. 2, 63
Weaver, J. M. 138
Webster, J. P. 13
Weismann, A. xii, xiv–xv, 2–4, 14, 18, 20–1, 24, 41–2, 54

Whitear, M. 38
Whitehead, A. N. 117–8
Wiener, A. S. 103
Williams, G. C. xii
Woglom, W. H. 10
Wong, H. 107
Woodger, J. H. 63
Wood-Jones, F. 72
wound healing 31, 110–4, 123–6
Wright, S. viii
wrinkles 36–7

X-rays 143–4

Zuckerman, S. 101

A CATALOGUE OF SELECTED DOVER BOOKS IN ALL FIELDS OF INTEREST

A CATALOGUE OF SELECTED DOVER BOOKS IN ALL FIELDS OF INTEREST

CONDITIONED REFLEXES, Ivan P. Pavlov. Full translation of most complete statement of Pavlov's work; cerebral damage, conditioned reflex, experiments with dogs, sleep, similar topics of great importance. 430pp. 5⅜ x 8½.
60614-7 Pa. $4.50

NOTES ON NURSING: WHAT IT IS, AND WHAT IT IS NOT, Florence Nightingale. Outspoken writings by founder of modern nursing. When first published (1860) it played an important role in much needed revolution in nursing. Still stimulating. 140pp. 5⅜ x 8½. 22340-X Pa. $2.50

HARTER'S PICTURE ARCHIVE FOR COLLAGE AND ILLUSTRATION, Jim Harter. Over 300 authentic, rare 19th-century engravings selected by noted collagist for artists, designers, decoupeurs, etc. Machines, people, animals, etc., printed one side of page. 25 scene plates for backgrounds. 6 collages by Harter, Satty, Singer, Evans. Introduction. 192pp. 8⅞ x 11¾.
23659-5 Pa. $5.00

MANUAL OF TRADITIONAL WOOD CARVING, edited by Paul N. Hasluck. Possibly the best book in English on the craft of wood carving. Practical instructions, along with 1,146 working drawings and photographic illustrations. Formerly titled *Cassell's Wood Carving*. 576pp. 6½ x 9¼.
23489-4 Pa. $7.95

THE PRINCIPLES AND PRACTICE OF HAND OR SIMPLE TURNING, John Jacob Holtzapffel. Full coverage of basic lathe techniques—history and development, special apparatus, softwood turning, hardwood turning, metal turning. Many projects—billiard ball, works formed within a sphere, egg cups, ash trays, vases, jardiniers, others—included. 1881 edition. 800 illustrations. 592pp. 6⅛ x 9¼. 23365-0 Clothbd. $15.00

THE JOY OF HANDWEAVING, Osma Tod. Only book you need for hand weaving. Fundamentals, threads, weaves, plus numerous projects for small board-loom, two-harness, tapestry, laid-in, four-harness weaving and more. Over 160 illustrations. 2nd revised edition. 352pp. 6½ x 9¼.
23458-4 Pa. $5.00

THE BOOK OF WOOD CARVING, Charles Marshall Sayers. Still finest book for beginning student in wood sculpture. Noted teacher, craftsman discusses fundamentals, technique; gives 34 designs, over 34 projects for panels, bookends, mirrors, etc. "Absolutely first-rate"—E. J. Tangerman. 33 photos. 118pp. 7¾ x 10⅝. 23654-4 Pa. $3.00

DRAWINGS OF WILLIAM BLAKE, William Blake. 92 plates from Book of Job, *Divine Comedy, Paradise Lost,* visionary heads, mythological figures, Laocoon, etc. Selection, introduction, commentary by Sir Geoffrey Keynes. 178pp. 8⅛ x 11. 22303-5 Pa. $4.00

ENGRAVINGS OF HOGARTH, William Hogarth. 101 of Hogarth's greatest works: *Rake's Progress, Harlot's Progress, Illustrations for Hudibras, Before and After, Beer Street and Gin Lane,* many more. Full commentary. 256pp. 11 x 13¾. 22479-1 Pa. $7.95

DAUMIER: 120 GREAT LITHOGRAPHS, Honore Daumier. Wide-ranging collection of lithographs by the greatest caricaturist of the 19th century. Concentrates on eternally popular series on lawyers, on married life, on liberated women, etc. Selection, introduction, and notes on plates by Charles F. Ramus. Total of 158pp. 9⅜ x 12¼. 23512-2 Pa. $5.50

DRAWINGS OF MUCHA, Alphonse Maria Mucha. Work reveals draftsman of highest caliber: studies for famous posters and paintings, renderings for book illustrations and ads, etc. 70 works, 9 in color; including 6 items not drawings. Introduction. List of illustrations. 72pp. 9⅜ x 12¼. (Available in U.S. only) 23672-2 Pa. $4.00

GIOVANNI BATTISTA PIRANESI: DRAWINGS IN THE PIERPONT MORGAN LIBRARY, Giovanni Battista Piranesi. For first time ever all of Morgan Library's collection, world's largest. 167 illustrations of rare Piranesi drawings—archeological, architectural, decorative and visionary. Essay, detailed list of drawings, chronology, captions. Edited by Felice Stampfle. 144pp. 9⅜ x 12¼. 23714-1 Pa. $7.50

NEW YORK ETCHINGS (1905-1949), John Sloan. All of important American artist's N.Y. life etchings. 67 works include some of his best art; also lively historical record—Greenwich Village, tenement scenes. Edited by Sloan's widow. Introduction and captions. 79pp. 8⅜ x 11¼. 23651-X Pa. $4.00

CHINESE PAINTING AND CALLIGRAPHY: A PICTORIAL SURVEY, Wan-go Weng. 69 fine examples from John M. Crawford's matchless private collection: landscapes, birds, flowers, human figures, etc., plus calligraphy. Every basic form included: hanging scrolls, handscrolls, album leaves, fans, etc. 109 illustrations. Introduction. Captions. 192pp. 8⅞ x 11¾. 23707-9 Pa. $7.95

DRAWINGS OF REMBRANDT, edited by Seymour Slive. Updated Lippmann, Hofstede de Groot edition, with definitive scholarly apparatus. All portraits, biblical sketches, landscapes, nudes, Oriental figures, classical studies, together with selection of work by followers. 550 illustrations. Total of 630pp. 9⅛ x 12¼. 21485-0, 21486-9 Pa., Two-vol. set $15.00

THE DISASTERS OF WAR, Francisco Goya. 83 etchings record horrors of Napoleonic wars in Spain and war in general. Reprint of 1st edition, plus 3 additional plates. Introduction by Philip Hofer. 97pp. 9⅜ x 8¼. 21872-4 Pa. $3.75

THE EARLY WORK OF AUBREY BEARDSLEY, Aubrey Beardsley. 157 plates, 2 in color: *Manon Lescaut, Madame Bovary, Morte Darthur, Salome,* other. Introduction by H. Marillier. 182pp. 8⅛ x 11. 21816-3 Pa. $4.50

THE LATER WORK OF AUBREY BEARDSLEY, Aubrey Beardsley. Exotic masterpieces of full maturity: *Venus and Tannhauser, Lysistrata, Rape of the Lock, Volpone,* Savoy material, etc. 174 plates, 2 in color. 186pp. 8⅛ x 11. 21817-1 Pa. $4.50

THOMAS NAST'S CHRISTMAS DRAWINGS, Thomas Nast. Almost all Christmas drawings by creator of image of Santa Claus as we know it, and one of America's foremost illustrators and political cartoonists. 66 illustrations. 3 illustrations in color on covers. 96pp. 8⅜ x 11¼. 23660-9 Pa. $3.50

THE DORÉ ILLUSTRATIONS FOR DANTE'S DIVINE COMEDY, Gustave Doré. All 135 plates from Inferno, Purgatory, Paradise; fantastic tortures, infernal landscapes, celestial wonders. Each plate with appropriate (translated) verses. 141pp. 9 x 12. 23231-X Pa. $4.50

DORÉ'S ILLUSTRATIONS FOR RABELAIS, Gustave Doré. 252 striking illustrations of *Gargantua and Pantagruel* books by foremost 19th-century illustrator. Including 60 plates, 192 delightful smaller illustrations. 153pp. 9 x 12. 23656-0 Pa. $5.00

LONDON: A PILGRIMAGE, Gustave Doré, Blanchard Jerrold. Squalor, riches, misery, beauty of mid-Victorian metropolis; 55 wonderful plates, 125 other illustrations, full social, cultural text by Jerrold. 191pp. of text. 9⅜ x 12¼. 22306-X Pa. $6.00

THE RIME OF THE ANCIENT MARINER, Gustave Doré, S. T. Coleridge. Dore's finest work, 34 plates capture moods, subtleties of poem. Full text. Introduction by Millicent Rose. 77pp. 9¼ x 12. 22305-1 Pa. $3.50

THE DORE BIBLE ILLUSTRATIONS, Gustave Doré. All wonderful, detailed plates: Adam and Eve, Flood, Babylon, Life of Jesus, etc. Brief King James text with each plate. Introduction by Millicent Rose. 241 plates. 241pp. 9 x 12. 23004-X Pa. $6.00

THE COMPLETE ENGRAVINGS, ETCHINGS AND DRYPOINTS OF ALBRECHT DURER. "Knight, Death and Devil"; "Melencolia," and more—all Dürer's known works in all three media, including 6 works formerly attributed to him. 120 plates. 235pp. 8⅜ x 11¼. 22851-7 Pa. $6.50

MAXIMILIAN'S TRIUMPHAL ARCH, Albrecht Dürer and others. Incredible monument of woodcut art: 8 foot high elaborate arch—heraldic figures, humans, battle scenes, fantastic elements—that you can assemble yourself. Printed on one side, layout for assembly. 143pp. 11 x 16. 21451-6 Pa. $5.00

THE COMPLETE WOODCUTS OF ALBRECHT DURER, edited by Dr. W. Kurth. 346 in all: "Old Testament," "St. Jerome," "Passion," "Life of Virgin," Apocalypse," many others. Introduction by Campbell Dodgson. 285pp. 8½ x 12¼. 21097-9 Pa. $6.95

DRAWINGS OF ALBRECHT DURER, edited by Heinrich Wolfflin. 81 plates show development from youth to full style. Many favorites; many new. Introduction by Alfred Werner. 96pp. 8⅛ x 11. 22352-3 Pa. $5.00

THE HUMAN FIGURE, Albrecht Dürer. Experiments in various techniques—stereometric, progressive proportional, and others. Also life studies that rank among finest ever done. Complete reprinting of *Dresden Sketchbook*. 170 plates. 355pp. 8⅜ x 11¼. 21042-1 Pa. $7.95

OF THE JUST SHAPING OF LETTERS, Albrecht Dürer. Renaissance artist explains design of Roman majuscules by geometry, also Gothic lower and capitals. Grolier Club edition. 43pp. 7⅞ x 10¾ 21306-4 Pa. $3.00

TEN BOOKS ON ARCHITECTURE, Vitruvius. The most important book ever written on architecture. Early Roman aesthetics, technology, classical orders, site selection, all other aspects. Stands behind everything since. Morgan translation. 331pp. 5⅜ x 8½. 20645-9 Pa. $4.00

THE FOUR BOOKS OF ARCHITECTURE, Andrea Palladio. 16th-century classic responsible for Palladian movement and style. Covers classical architectural remains, Renaissance revivals, classical orders, etc. 1738 Ware English edition. Introduction by A. Placzek. 216 plates. 110pp. of text. 9½ x 12¾. 21308-0 Pa. $8.95

HORIZONS, Norman Bel Geddes. Great industrialist stage designer, "father of streamlining," on application of aesthetics to transportation, amusement, architecture, etc. 1932 prophetic account; function, theory, specific projects. 222 illustrations. 312pp. 7⅞ x 10¾. 23514-9 Pa. $6.95

FRANK LLOYD WRIGHT'S FALLINGWATER, Donald Hoffmann. Full, illustrated story of conception and building of Wright's masterwork at Bear Run, Pa. 100 photographs of site, construction, and details of completed structure. 112pp. 9¼ x 10. 23671-4 Pa. $5.00

THE ELEMENTS OF DRAWING, John Ruskin. Timeless classic by great Viltorian; starts with basic ideas, works through more difficult. Many practical exercises. 48 illustrations. Introduction by Lawrence Campbell. 228pp. 5⅜ x 8½. 22730-8 Pa. $2.75

GIST OF ART, John Sloan. Greatest modern American teacher, Art Students League, offers innumerable hints, instructions, guided comments to help you in painting. Not a formal course. 46 illustrations. Introduction by Helen Sloan. 200pp. 5⅜ x 8½. 23435-5 Pa. $3.50

THE ANATOMY OF THE HORSE, George Stubbs. Often considered the great masterpiece of animal anatomy. Full reproduction of 1766 edition, plus prospectus; original text and modernized text. 36 plates. Introduction by Eleanor Garvey. 121pp. 11 x 14¾. 23402-9 Pa. $6.00

BRIDGMAN'S LIFE DRAWING, George B. Bridgman. More than 500 illustrative drawings and text teach you to abstract the body into its major masses, use light and shade, proportion; as well as specific areas of anatomy, of which Bridgman is master. 192pp. 6½ x 9¼. (Available in U.S. only) 22710-3 Pa. $3.00

ART NOUVEAU DESIGNS IN COLOR, Alphonse Mucha, Maurice Verneuil, Georges Auriol. Full-color reproduction of *Combinaisons ornementales* (c. 1900) by Art Nouveau masters. Floral, animal, geometric, interlacings, swashes—borders, frames, spots—all incredibly beautiful. 60 plates, hundreds of designs. 9⅜ x 8-1/16. 22885-1 Pa. $4.00

FULL-COLOR FLORAL DESIGNS IN THE ART NOUVEAU STYLE, E. A. Seguy. 166 motifs, on 40 plates, from *Les fleurs et leurs applications decoratives* (1902): borders, circular designs, repeats, allovers, "spots." All in authentic Art Nouveau colors. 48pp. 9⅜ x 12¼. 23439-8 Pa. $5.00

A DIDEROT PICTORIAL ENCYCLOPEDIA OF TRADES AND INDUSTRY, edited by Charles C. Gillispie. 485 most interesting plates from the great French Encyclopedia of the 18th century show hundreds of working figures, artifacts, process, land and cityscapes; glassmaking, papermaking, metal extraction, construction, weaving, making furniture, clothing, wigs, dozens of other activities. Plates fully explained. 920pp. 9 x 12. 22284-5, 22285-3 Clothbd., Two-vol. set $40.00

HANDBOOK OF EARLY ADVERTISING ART, Clarence P. Hornung. Largest collection of copyright-free early and antique advertising art ever compiled. Over 6,000 illustrations, from Franklin's time to the 1890's for special effects, novelty. Valuable source, almost inexhaustible.
Pictorial Volume. Agriculture, the zodiac, animals, autos, birds, Christmas, fire engines, flowers, trees, musical instruments, ships, games and sports, much more. Arranged by subject matter and use. 237 plates. 288pp. 9 x 12. 20122-8 Clothbd. $13.50

Typographical Volume. Roman and Gothic faces ranging from 10 point to 300 point, "Barnum," German and Old English faces, script, logotypes, scrolls and flourishes, 1115 ornamental initials, 67 complete alphabets, more. 310 plates. 320pp. 9 x 12. 20123-6 Clothbd. $15.00

CALLIGRAPHY (CALLIGRAPHIA LATINA), J. G. Schwandner. High point of 18th-century ornamental calligraphy. Very ornate initials, scrolls, borders, cherubs, birds, lettered examples. 172pp. 9 x 13. 20475-8 Pa. $6.00

ART FORMS IN NATURE, Ernst Haeckel. Multitude of strangely beautiful natural forms: Radiolaria, Foraminifera, jellyfishes, fungi, turtles, bats, etc. All 100 plates of the 19th-century evolutionist's *Kunstformen der Natur* (1904). 100pp. 9⅜ x 12¼. 22987-4 Pa. $4.50

CHILDREN: A PICTORIAL ARCHIVE FROM NINETEENTH-CENTURY SOURCES, edited by Carol Belanger Grafton. 242 rare, copyright-free wood engravings for artists and designers. Widest such selection available. All illustrations in line. 119pp. 8⅜ x 11¼. 23694-3 Pa. $3.50

WOMEN: A PICTORIAL ARCHIVE FROM NINETEENTH-CENTURY SOURCES, edited by Jim Harter. 391 copyright-free wood engravings for artists and designers selected from rare periodicals. Most extensive such collection available. All illustrations in line. 128pp. 9 x 12. 23703-6 Pa. $4.50

ARABIC ART IN COLOR, Prisse d'Avennes. From the greatest ornamentalists of all time—50 plates in color, rarely seen outside the Near East, rich in suggestion and stimulus. Includes 4 plates on covers. 46pp. 9⅜ x 12¼. 23658-7 Pa. $6.00

AUTHENTIC ALGERIAN CARPET DESIGNS AND MOTIFS, edited by June Beveridge. Algerian carpets are world famous. Dozens of geometrical motifs are charted on grids, color-coded, for weavers, needleworkers, craftsmen, designers. 53 illustrations plus 4 in color. 48pp. 8¼ x 11. (Available in U.S. only) 23650-1 Pa. $1.75

DICTIONARY OF AMERICAN PORTRAITS, edited by Hayward and Blanche Cirker. 4000 important Americans, earliest times to 1905, mostly in clear line. Politicians, writers, soldiers, scientists, inventors, industrialists, Indians, Blacks, women, outlaws, etc. Identificatory information. 756pp. 9¼ x 12¾. 21823-6 Clothbd. $40.00

HOW THE OTHER HALF LIVES, Jacob A. Riis. Journalistic record of filth, degradation, upward drive in New York immigrant slums, shops, around 1900. New edition includes 100 original Riis photos, monuments of early photography. 233pp. 10 x 7⅞. 22012-5 Pa. $6.00

NEW YORK IN THE THIRTIES, Berenice Abbott. Noted photographer's fascinating study of city shows new buildings that have become famous and old sights that have disappeared forever. Insightful commentary. 97 photographs. 97pp. 11⅜ x 10. 22967-X Pa. $5.00

MEN AT WORK, Lewis W. Hine. Famous photographic studies of construction workers, railroad men, factory workers and coal miners. New supplement of 18 photos on Empire State building construction. New introduction by Jonathan L. Doherty. Total of 69 photos. 63pp. 8 x 10¾. 23475-4 Pa. $3.00

THE DEPRESSION YEARS AS PHOTOGRAPHED BY ARTHUR ROTHSTEIN, Arthur Rothstein. First collection devoted entirely to the work of outstanding 1930s photographer: famous dust storm photo, ragged children, unemployed, etc. 120 photographs. Captions. 119pp. 9¼ x 10¾.
23590-4 Pa. $5.00

CAMERA WORK: A PICTORIAL GUIDE, Alfred Stieglitz. All 559 illustrations and plates from the most important periodical in the history of art photography, *Camera Work* (1903-17). Presented four to a page, reduced in size but still clear, in strict chronological order, with complete captions. Three indexes. Glossary. Bibliography. 176pp. 8⅜ x 11¼.
23591-2 Pa. $6.95

ALVIN LANGDON COBURN, PHOTOGRAPHER, Alvin L. Coburn. Revealing autobiography by one of greatest photographers of 20th century gives insider's version of Photo-Secession, plus comments on his own work. 77 photographs by Coburn. Edited by Helmut and Alison Gernsheim. 160pp. 8⅛ x 11.
23685-4 Pa. $6.00

NEW YORK IN THE FORTIES, Andreas Feininger. 162 brilliant photographs by the well-known photographer, formerly with *Life* magazine, show commuters, shoppers, Times Square at night, Harlem nightclub, Lower East Side, etc. Introduction and full captions by John von Hartz. 181pp. 9¼ x 10¾.
23585-8 Pa. $6.00

GREAT NEWS PHOTOS AND THE STORIES BEHIND THEM, John Faber. Dramatic volume of 140 great news photos, 1855 through 1976, and revealing stories behind them, with both historical and technical information. Hindenburg disaster, shooting of Oswald, nomination of Jimmy Carter, etc. 160pp. 8¼ x 11.
23667-6 Pa. $5.00

THE ART OF THE CINEMATOGRAPHER, Leonard Maltin. Survey of American cinematography history and anecdotal interviews with 5 masters—Arthur Miller, Hal Mohr, Hal Rosson, Lucien Ballard, and Conrad Hall. Very large selection of behind-the-scenes production photos. 105 photographs. Filmographies. Index. Originally *Behind the Camera.* 144pp. 8¼ x 11.
23686-2 Pa. $5.00

DESIGNS FOR THE THREE-CORNERED HAT (LE TRICORNE), Pablo Picasso. 32 fabulously rare drawings—including 31 color illustrations of costumes and accessories—for 1919 production of famous ballet. Edited by Parmenia Migel, who has written new introduction. 48pp. 9⅜ x 12¼. (Available in U.S. only)
23709-5 Pa. $5.00

NOTES OF A FILM DIRECTOR, Sergei Eisenstein. Greatest Russian filmmaker explains montage, making of *Alexander Nevsky,* aesthetics; comments on self, associates, great rivals (Chaplin), similar material. 78 illustrations. 240pp. 5⅜ x 8½.
22392-2 Pa. $4.50

HOLLYWOOD GLAMOUR PORTRAITS, edited by John Kobal. 145 photos capture the stars from 1926-49, the high point in portrait photography. Gable, Harlow, Bogart, Bacall, Hedy Lamarr, Marlene Dietrich, Robert Montgomery, Marlon Brando, Veronica Lake; 94 stars in all. Full background on photographers, technical aspects, much more. Total of 160pp. 8⅜ x 11¼. 23352-9 Pa. $5.00

THE NEW YORK STAGE: FAMOUS PRODUCTIONS IN PHOTOGRAPHS, edited by Stanley Appelbaum. 148 photographs from Museum of City of New York show 142 plays, 1883-1939. *Peter Pan, The Front Page, Dead End, Our Town,* O'Neill, hundreds of actors and actresses, etc. Full indexes. 154pp. 9½ x 10. 23241-7 Pa. $6.00

MASTERS OF THE DRAMA, John Gassner. Most comprehensive history of the drama, every tradition from Greeks to modern Europe and America, including Orient. Covers 800 dramatists, 2000 plays; biography, plot summaries, criticism, theatre history, etc. 77 illustrations. 890pp. 5⅜ x 8½. 20100-7 Clothbd. $10.00

THE GREAT OPERA STARS IN HISTORIC PHOTOGRAPHS, edited by James Camner. 343 portraits from the 1850s to the 1940s: Tamburini, Mario, Caliapin, Jeritza, Melchior, Melba, Patti, Pinza, Schipa, Caruso, Farrar, Steber, Gobbi, and many more—270 performers in all. Index. 199pp. 8⅜ x 11¼. 23575-0 Pa. $6.50

J. S. BACH, Albert Schweitzer. Great full-length study of Bach, life, background to music, music, by foremost modern scholar. Ernest Newman translation. 650 musical examples. Total of 928pp. 5⅜ x 8½. (Available in U.S. only) 21631-4, 21632-2 Pa., Two-vol. set $10.00

COMPLETE PIANO SONATAS, Ludwig van Beethoven. All sonatas in the fine Schenker edition, with fingering, analytical material. One of best modern editions. Total of 615pp. 9 x 12. (Available in U.S. only) 23134-8, 23135-6 Pa., Two-vol. set $15.00

KEYBOARD MUSIC, J. S. Bach. Bach-Gesellschaft edition. For harpsichord, piano, other keyboard instruments. English Suites, French Suites, Six Partitas, Goldberg Variations, Two-Part Inventions, Three-Part Sinfonias. 312pp. 8⅛ x 11. (Available in U.S. only) 22360-4 Pa. $6.00

FOUR SYMPHONIES IN FULL SCORE, Franz Schubert. Schubert's four most popular symphonies: No. 4 in C Minor ("Tragic"); No. 5 in B-flat Major; No. 8 in B Minor ("Unfinished"); No. 9 in C Major ("Great"). Breitkopf & Hartel edition. Study score. 261pp. 9⅜ x 12¼. 23681-1 Pa. $6.50

THE AUTHENTIC GILBERT & SULLIVAN SONGBOOK, W. S. Gilbert, A. S. Sullivan. Largest selection available; 92 songs, uncut, original keys, in piano rendering approved by Sullivan. Favorites and lesser-known fine numbers. Edited with plot synopses by James Spero. 3 illustrations. 399pp. 9 x 12. 23482-7 Pa. $7.95

PRINCIPLES OF ORCHESTRATION, Nikolay Rimsky-Korsakov. Great classical orchestrator provides fundamentals of tonal resonance, progression of parts, voice and orchestra, tutti effects, much else in major document. 330pp. of musical excerpts. 489pp. 6½ x 9¼. 21266-1 Pa. $6.00

TRISTAN UND ISOLDE, Richard Wagner. Full orchestral score with complete instrumentation. Do not confuse with piano reduction. Commentary by Felix Mottl, great Wagnerian conductor and scholar. Study score. 655pp. 8⅛ x 11. 22915-7 Pa. $12.50

REQUIEM IN FULL SCORE, Giuseppe Verdi. Immensely popular with choral groups and music lovers. Republication of edition published by C. F. Peters, Leipzig, n. d. German frontmaker in English translation. Glossary. Text in Latin. Study score. 204pp. 9⅜ x 12¼. 23682-X Pa. $6.00

COMPLETE CHAMBER MUSIC FOR STRINGS, Felix Mendelssohn. All of Mendelssohn's chamber music: Octet, 2 Quintets, 6 Quartets, and Four Pieces for String Quartet. (Nothing with piano is included). Complete works edition (1874-7). Study score. 283 pp. 9⅜ x 12¼. 23679-X Pa. $6.95

POPULAR SONGS OF NINETEENTH-CENTURY AMERICA, edited by Richard Jackson. 64 most important songs: "Old Oaken Bucket," "Arkansas Traveler," "Yellow Rose of Texas," etc. Authentic original sheet music, full introduction and commentaries. 290pp. 9 x 12. 23270-0 Pa. $6.00

COLLECTED PIANO WORKS, Scott Joplin. Edited by Vera Brodsky Lawrence. Practically all of Joplin's piano works—rags, two-steps, marches, waltzes, etc., 51 works in all. Extensive introduction by Rudi Blesh. Total of 345pp. 9 x 12. 23106-2 Pa. $14.95

BASIC PRINCIPLES OF CLASSICAL BALLET, Agrippina Vaganova. Great Russian theoretician, teacher explains methods for teaching classical ballet; incorporates best from French, Italian, Russian schools. 118 illustrations. 175pp. 5⅜ x 8½. 22036-2 Pa. $2.50

CHINESE CHARACTERS, L. Wieger. Rich analysis of 2300 characters according to traditional systems into primitives. Historical-semantic analysis to phonetics (Classical Mandarin) and radicals. 820pp. 6⅛ x 9¼. 21321-8 Pa. $10.00

EGYPTIAN LANGUAGE: EASY LESSONS IN EGYPTIAN HIEROGLYPHICS, E. A. Wallis Budge. Foremost Egyptologist offers Egyptian grammar, explanation of hieroglyphics, many reading texts, dictionary of symbols. 246pp. 5 x 7½. (Available in U.S. only) 21394-3 Clothbd. $7.50

AN ETYMOLOGICAL DICTIONARY OF MODERN ENGLISH, Ernest Weekley. Richest, fullest work, by foremost British lexicographer. Detailed word histories. Inexhaustible. Do not confuse this with *Concise Etymological Dictionary*, which is abridged. Total of 856pp. 6½ x 9¼. 21873-2, 21874-0 Pa., Two-vol. set $12.00

A MAYA GRAMMAR, Alfred M. Tozzer. Practical, useful English-language grammar by the Harvard anthropologist who was one of the three greatest American scholars in the area of Maya culture. Phonetics, grammatical processes, syntax, more. 301pp. 5⅜ x 8½. 23465-7 Pa. $4.00

THE JOURNAL OF HENRY D. THOREAU, edited by Bradford Torrey, F. H. Allen. Complete reprinting of 14 volumes, 1837-61, over two million words; the sourcebooks for *Walden*, etc. Definitive. All original sketches, plus 75 photographs. Introduction by Walter Harding. Total of 1804pp. 8½ x 12¼. 20312-3, 20313-1 Clothbd., Two-vol. set $50.00

CLASSIC GHOST STORIES, Charles Dickens and others. 18 wonderful stories you've wanted to reread: "The Monkey's Paw," "The House and the Brain," "The Upper Berth," "The Signalman," "Dracula's Guest," "The Tapestried Chamber," etc. Dickens, Scott, Mary Shelley, Stoker, etc. 330pp. 5⅜ x 8½. 20735-8 Pa. $3.50

SEVEN SCIENCE FICTION NOVELS, H. G. Wells. Full novels. *First Men in the Moon, Island of Dr. Moreau, War of the Worlds, Food of the Gods, Invisible Man, Time Machine, In the Days of the Comet*. A basic science-fiction library. 1015pp. 5⅜ x 8½. (Available in U.S. only) 20264-X Clothbd. $8.95

ARMADALE, Wilkie Collins. Third great mystery novel by the author of *The Woman in White* and *The Moonstone*. Ingeniously plotted narrative shows an exceptional command of character, incident and mood. Original magazine version with 40 illustrations. 597pp. 5⅜ x 8½. 23429-0 Pa. $5.00

MASTERS OF MYSTERY, H. Douglas Thomson. The first book in English (1931) devoted to history and aesthetics of detective story. Poe, Doyle, LeFanu, Dickens, many others, up to 1930. New introduction and notes by E. F. Bleiler. 288pp. 5⅜ x 8½. (Available in U.S. only) 23606-4 Pa. $4.00

FLATLAND, E. A. Abbott. Science-fiction classic explores life of 2-D being in 3-D world. Read also as introduction to thought about hyperspace. Introduction by Banesh Hoffmann. 16 illustrations. 103pp. 5⅜ x 8½. 20001-9 Pa. $1.75

THREE SUPERNATURAL NOVELS OF THE VICTORIAN PERIOD, edited, with an introduction, by E. F. Bleiler. Reprinted complete and unabridged, three great classics of the supernatural: *The Haunted Hotel* by Wilkie Collins, *The Haunted House at Latchford* by Mrs. J. H. Riddell, and *The Lost Stradivarious* by J. Meade Falkner. 325pp. 5⅜ x 8½. 22571-2 Pa. $4.00

AYESHA: THE RETURN OF "SHE," H. Rider Haggard. Virtuoso sequel featuring the great mythic creation, Ayesha, in an adventure that is fully as good as the first book, *She*. Original magazine version, with 47 original illustrations by Maurice Greiffenhagen. 189pp. 6½ x 9¼. 23649-8 Pa. $3.50

UNCLE SILAS, J. Sheridan LeFanu. Victorian Gothic mystery novel, considered by many best of period, even better than Collins or Dickens. Wonderful psychological terror. Introduction by Frederick Shroyer. 436pp. 5⅜ x 8½. 21715-9 Pa. $6.00

JURGEN, James Branch Cabell. The great erotic fantasy of the 1920's that delighted thousands, shocked thousands more. Full final text, Lane edition with 13 plates by Frank Pape. 346pp. 5⅜ x 8½. 23507-6 Pa. $4.50

THE CLAVERINGS, Anthony Trollope. Major novel, chronicling aspects of British Victorian society, personalities. Reprint of Cornhill serialization, 16 plates by M. Edwards; first reprint of full text. Introduction by Norman Donaldson. 412pp. 5⅜ x 8½. 23464-9 Pa. $5.00

KEPT IN THE DARK, Anthony Trollope. Unusual short novel about Victorian morality and abnormal psychology by the great English author. Probably the first American publication. Frontispiece by Sir John Millais. 92pp. 6½ x 9¼. 23609-9 Pa. $2.50

RALPH THE HEIR, Anthony Trollope. Forgotten tale of illegitimacy, inheritance. Master novel of Trollope's later years. Victorian country estates, clubs, Parliament, fox hunting, world of fully realized characters. Reprint of 1871 edition. 12 illustrations by F. A. Faser. 434pp. of text. 5⅜ x 8½. 23642-0 Pa. $5.00

YEKL and THE IMPORTED BRIDEGROOM AND OTHER STORIES OF THE NEW YORK GHETTO, Abraham Cahan. Film *Hester Street* based on *Yekl* (1896). Novel, other stories among first about Jewish immigrants of N.Y.'s East Side. Highly praised by W. D. Howells—Cahan "a new star of realism." New introduction by Bernard G. Richards. 240pp. 5⅜ x 8½. 22427-9 Pa. $3.50

THE HIGH PLACE, James Branch Cabell. Great fantasy writer's enchanting comedy of disenchantment set in 18th-century France. Considered by some critics to be even better than his famous *Jurgen*. 10 illustrations and numerous vignettes by noted fantasy artist Frank C. Pape. 320pp. 5⅜ x 8½. 23670-6 Pa. $4.00

ALICE'S ADVENTURES UNDER GROUND, Lewis Carroll. Facsimile of ms. Carroll gave Alice Liddell in 1864. Different in many ways from final Alice. Handlettered, illustrated by Carroll. Introduction by Martin Gardner. 128pp. 5⅜ x 8½. 21482-6 Pa. $2.00

FAVORITE ANDREW LANG FAIRY TALE BOOKS IN MANY COLORS, Andrew Lang. The four Lang favorites in a boxed set—the complete *Red, Green, Yellow* and *Blue* Fairy Books. 164 stories; 439 illustrations by Lancelot Speed, Henry Ford and G. P. Jacomb Hood. Total of about 1500pp. 5⅜ x 8½. 23407-X Boxed set, Pa. $14.95

CATALOGUE OF BOOKS

HOUSEHOLD STORIES BY THE BROTHERS GRIMM. All the great Grimm stories: "Rumpelstiltskin," "Snow... "Hansel and Gretel," etc., with 114 illustrations by Walter Cra... 5⅜ x 8½.
21080-4 Pa. $3.00

SLEEPING BEAUTY, illustrated by Arthur R... Perhaps the fullest, most delightful version ever, told by C. S. E... ...kham's best work. 49 illustrations. 110pp. 7⅞ x 10¾.
22756-1 Pa. $2.50

AMERICAN FAIRY TALES, L. Frank Baum. ... cowboy lassoes Father Time; dummy in Mr. Floman's department s... ...indow comes to life; and 10 other fairy tales. 41 illustrations by N. P. ... Harry Kennedy, Ike Morgan, and Ralph Gardner. 209pp. 5⅜ x 8½.
...43-9 Pa. $3.00

THE WONDERFUL WIZARD OF OZ, L. Frank Baum. ...csimile in full color of America's finest children's classic. Introduction by ...rtin Gardner. 143 illustrations by W. W. Denslow. 267pp. 5⅜ x 8½.
2069... Pa. $3.50

THE TALE OF PETER RABBIT, Beatrix Potter. The inimitable Peter's terrifying adventure in Mr. McGregor's garden, with all 27 wonderful, full-color Potter illustrations. 55pp. 4¼ x 5½. (Available in U.S. only)
22827-4 Pa. $1.25

THE STORY OF KING ARTHUR AND HIS KNIGHTS, Howard Pyle. Finest children's version of life of King Arthur. 48 illustrations by Pyle. 131pp. 6⅛ x 9¼.
21445-1 Pa. $4.95

CARUSO'S CARICATURES, Enrico Caruso. Great tenor's remarkable caricatures of self, fellow musicians, composers, others. Toscanini, Puccini, Farrar, etc. Impish, cutting, insightful. 473 illustrations. Preface by M. Sisca. 217pp. 8⅜ x 11¼.
23528-9 Pa. $6.95

PERSONAL NARRATIVE OF A PILGRIMAGE TO ALMADINAH AND MECCAH, Richard Burton. Great travel classic by remarkably colorful personality. Burton, disguised as a Moroccan, visited sacred shrines of Islam, narrowly escaping death. Wonderful observations of Islamic life, customs, personalities. 47 illustrations. Total of 959pp. 5⅜ x 8½.
21217-3, 21218-1 Pa., Two-vol. set $12.00

INCIDENTS OF TRAVEL IN YUCATAN, John L. Stephens. Classic (1843) exploration of jungles of Yucatan, looking for evidences of Maya civilization. Travel adventures, Mexican and Indian culture, etc. Total of 669pp. 5⅜ x 8½.
20926-1, 20927-X Pa., Two-vol. set $7.90

AMERICAN LITERARY AUTOGRAPHS FROM WASHINGTON IRVING TO HENRY JAMES, Herbert Cahoon, et al. Letters, poems, manuscripts of Hawthorne, Thoreau, Twain, Alcott, Whitman, 67 other prominent American authors. Reproductions, full transcripts and commentary. Plus checklist of all American Literary Autographs in The Pierpont Morgan Library. Printed on exceptionally high-quality paper. 136 illustrations. 212pp. 9⅛ x 12¼.
23548-3 Pa. $7.95

CATALOGU R BOOKS

YUCATAN BEFORE AND CONQUEST, Diego de Landa. First English translation of Maya studies, the only significant account of Yucatan writt y post-Conquest era. Translated by distinguished Maya scho Gates. Appendices, introduction, 4 maps and over 120 illu ed by translator. 162pp. 5⅜ x 8½.
23622-6 Pa. $3.00

THE MALAY ARCH), Alfred R. Wallace. Spirited travel account by one of founders o biology. Touches on zoology, botany, ethnography, geography, logy. 62 illustrations, maps. 515pp. 5⅜ x 8½.
20187-2 Pa. $6.95

THE DISCOVE OF THE TOMB OF TUTANKHAMEN, Howard Carter, A. C. M Accompany Carter in the thrill of discovery, as ruined passage sudde reveals unique, untouched, fabulously rich tomb. Fascinating accoun with 106 illustrations. New introduction by J. M. White. Total of 382p. 5⅜ x 8½. (Available in U.S. only) 23500-9 Pa. $4.00

THE WORLD'S GREATEST SPEECHES, edited by Lewis Copeland and Lawrence W. Lamm. Vast collection of 278 speeches from Greeks up to present. Powerful and effective models; unique look at history. Revised to 1970. Indices. 842pp. 5⅜ x 8½.
20468-5 Pa. $8.95

THE 100 GREATEST ADVERTISEMENTS, Julian Watkins. The priceless ingredient; His master's voice; 99 44/100% pure; over 100 others. How they were written, their impact, etc. Remarkable record. 130 illustrations. 233pp. 7⅞ x 10 3/5.
20540-1 Pa. $5.00

CRUICKSHANK PRINTS FOR HAND COLORING, George Cruickshank. 18 illustrations, one side of a page, on fine-quality paper suitable for watercolors. Caricatures of people in society (c. 1820) full of trenchant wit. Very large format. 32pp. 11 x 16.
23684-6 Pa. $5.00

THIRTY-TWO COLOR POSTCARDS OF TWENTIETH-CENTURY AMERICAN ART, Whitney Museum of American Art. Reproduced in full color in postcard form are 31 art works and one shot of the museum. Calder, Hopper, Rauschenberg, others. Detachable. 16pp. 8¼ x 11.
23629-3 Pa. $2.50

MUSIC OF THE SPHERES: THE MATERIAL UNIVERSE FROM ATOM TO QUASAR SIMPLY EXPLAINED, Guy Murchie. Planets, stars, geology, atoms, radiation, relativity, quantum theory, light, antimatter, similar topics. 319 figures. 664pp. 5⅜ x 8½.
21809-0, 21810-4 Pa., Two-vol. set $10.00

EINSTEIN'S THEORY OF RELATIVITY, Max Born. Finest semi-technical account; covers Einstein, Lorentz, Minkowski, and others, with much detail, much explanation of ideas and math not readily available elsewhere on this level. For student, non-specialist. 376pp. 5⅜ x 8½.
60769-0 Pa. $4.50